LONE PINE

Reptiles *of the* Northwest

Text, photographs, and drawings by
Alan St. John

First printed in 2002 10 9 8 7 6 5 4 3
Printed in China

The Publisher: Lone Pine Publishing

1808 B Street NW, Suite 140
Auburn, WA 98001
U.S.A.

10145 – 81 Avenue
Edmonton, Alberta T6E 1W9
Canada

Website: http://www.lonepinepublishing.com

National Library of Canada Cataloguing in Publication Data

St. John, Alan D.

Reptiles of the Northwest: Alaska to California; Rockies to the Coast

Includes bibliographical references and index.
ISBN 1-55105-343-8

Also published as Reptiles of the Northwest: British Columbia to California (1-55105-349-7); C2002-910425-4

Reptiles—Northwest, Pacific. I. Title.
QL653.N75S26 2002 597.9'09795 C2002-910397-5

Editorial Director: Nancy Foulds
Project Editor: Roland Lines
Editorial: Roland Lines, Genevieve Boyer
Production Co-ordinator: Jennifer Fafard
Cover Design: Elliot Engley, Rod Michalchuk
Book Design: Elliot Engley
Photography & Illustrations: Alan St. John
Cartography: Elliot Engley
Separations & Film: Elite Lithography

We acknowledge the financial support of the Government of Canada through the Book Publishing Industry Development Program (BPIDP) for our publishing activities.

PC: P13

I dedicate this book to young herpetologists and naturalists—may their tribe increase—and to mentor, friend, and elder of the tribe, Robert "Doc" M. Storm, Emeritus Professor of Zoology, Oregon State University. My first meeting with Doc Storm occurred when I was 15 years old, at which time I showed him a garter snake that I couldn't identify and peppered him with a barrage of herpetological questions. He patiently answered all my queries and has continued to generously encourage and inspire me all these years. Thanks, Doc!

Doc Storm at Pyramid Lake, northwestern Nevada.

Acknowledgments

I sincerely wish to convey to all the kind people who helped me with the creation of this book how much their time and trouble has been appreciated. Faithful friends and family frequently dropped whatever they were doing to aid my search for certain elusive reptiles. The memories of happy times spent together flipping rocks, noosing lizards, road hunting, and sharing nights around sagebrush campfires will always be treasured.

In my years of travel throughout the Northwest I have met many wonderful people and gained a number of new friends. They enthusiastically shared information, acted as guides to hard-to-find places, caught specimens, loaned me prized pet snakes, helped while I photographed uncooperative reptiles, and fed and housed me on occasion. Others spent considerable time answering my questions during phone conversations or sent their unpublished research for me to delve into. Sometimes I despaired that I would ever capture and photograph all the reptiles required. The task would not have been accomplished without the various contributions of so many people.

Foremost, I would like to express my deepest gratitude to Buck Jenkins, supporter of the arts, literature, and environmental conservation education and all around enthusiastic connoisseur of the world of nature. Without his generous funding for most of the research, travel expenses, and supplies that went into the production of this field guide, the project would have been literally impossible.

Thanks also go to Lone Pine Publishing, particularly editors Nancy Foulds and Roland Lines. Their genuine interest in what I wanted to accomplish in this book and professional expertise in bringing it to fruition has been greatly appreciated. Lone Pine's commitment to producing a truly comprehensive series of durable, useful field guides is admirable. It is a boon to all naturalists that Lone Pine has gone beyond the more popular wildflower and bird titles to also include subjects like fungi and "creepy-crawly" amphibians and reptiles.

I am grateful to the following people who read through all or parts of the early drafts of the text and gave helpful suggestions and encouragement that significantly improved the book: John Applegarth, Breck Bartholomew, Edmund Brodie Jr., Doug Calvin, Jim England, Bob Espinoza, Dorothy Fender, Dan Holland, Richard Hoyer, Buck Jenkins, Jimmy Kagan, Doug Knutsen, Bill Leonard, Kelly McAllister, Jim McMahon, Martin Nugent, Gayle Parlato, Chris Pearl, Pete and Gretchen Pederson, Lee Simons, Hobart Smith, Robert Stebbins, Bob Storm, Bruce Taylor, Tom Titus, and Craig Zuger.

Those who supplied locality data or taxonomic clarifications, shared their field observations on certain species, dug into past records, extracted old journal entries, and helped in numerous other ways are Marvin Abst, Ed Alverson, Jim Anderson, Ralph Anderson, Reuben Anderson, John Applegarth, John Arnold, Steve Arnold, Jen Ballard, Breck Bartholomew, Jim Baugh, Ken Beatty, Mary Benterou, Sue Bielke, Rick Blair, Marty Brae, Edmund Brodie Jr., Sharon Browder, Paul Brown, Rodger Bryan, Bruce Bury, Doug Calvin, Mark Caddy, Mark Chatigny, Tim Clarke,

Dave Clayton, Vic Coggins, Char Corkran, Steve Corn, Stewart Croghan, Steven Cross, Sam Cuenca, Dave Danley, Jim David, Keith Day, Teresa DeLorenzo, Rita Dixon, Robin Dobson, David Doty, Jim Douglas, Mary Jo Douglas, Mike Dunkelberger, Frank Eddleman, Jack Eddy, Christian Engelstoft, Jim England, Bob Espinoza, Terry Farrell, Roger Farschon, Dennis Flath, Nora Foster, Terrence Frest, Laura Friis, Ted Gahr, Eleanor Gaines, Larry Gangle, Rebecca Goggins, Harry Greene, Pat Gregory, Geoffrey Hammerson, Chuck Harris, Marc Hayes, Jim Heinrichs, Calvin Henry, Pete Hill, Dan Holland, Carl Holte, Joann Homuth, Richard Hoyer, Brian Hubbs, Peter Hujik, Jim and Celia Huygens, Pat Jamieson, Buck Jenkins, Bruce Johnson, Mike Judd, Jackie Kephart, Arne Knutsen, Doug Knutsen, Hans Konig, Mike Lais, Chris Lapp, Bob Larsen, John Legler, Bill Leonard, Susan Linsted, Erik Lyons, Bryce Maxell, Kelly McAllister, Scott McElveen, Sara McGuire, Jim McMahon, Ron Mechlinberg, Laura Meir, Larry Michaels, Jerry Mires, Bryce Mitchell, Ron Morton, Kit Novick, Ron Nussbaum, George Oliver, Kristiina Ovaska, Ron Panik, Ed Park, Tim Patton, Brad Petch, Steve Petersburg, Brian Paust, Chris Pearl, Chuck Peterson, Mike Pfrender, Susan Piper, Roy Price, Mary Rasmussen, Jim Riggs, Greg Robart, Jan Roth, Jon Sadowski, Jerod Sapp, Mike Sarrel, Lee Simons, Jack Sites, Eric Skov, Frank Slavins, Hobart Smith, Regina Smith, Rick Staub, Robert Stebbins, Scott Stevens, Elaine Stewart, Bob Storm, Lois Stuart, John Surgenor, Karl Switak, Cynthia Tait, Wilmer Tanner, Larry Teske, Rich Thurman, Tom Titus, Jim Torland, Ron Toroni, Ann Trieu, Natalie Turley, Dick VanderSchaff, Jens Vindum, Randy Webb, John Wells, Hart Welsh, Kerwin Werner, Neal Young, Steve Zachary, Kelly Zamudeo, and Craig Zuger.

I would also like to express my gratitude to the following institutions and their staff representatives who supplied me with excerpts from their herpetological records and/or loaned me preserved specimens for examination: Brigham Young University (Jack Sites), California Academy of Sciences (Jens Vindum), Harvard Museum of Comparative Zoology (Jose Rosado), Oregon State University (Doug Markle), Shasta College (Marvin Abst), University of California, Berkeley (Carla Cicero), University of Nevada, Reno (Bob Espinoza), University of Utah (Eric Rickart), University of Wyoming (Tim Patton), and Western Nevada Community College (Ron Panik). A special thanks goes to Mark Eberle and the Central Oregon Community College Science Department, which acted as an institutional interface in the loan of several preserved specimens.

Live specimens for photographing were provided by Rick Blair, Brad's World of Reptiles (Brad Tylman and Mark Mensch), Wade Combs, Greg Conaway, John Cossel Jr., Mike Dunkelberger, Flying M Ranch (Summer Sequeira), Steve Harrison, The High Desert Museum (Becky Anderson and Kelly Schaub), Richard Hoyer, Amy Jacobs, Doug Knutsen, Steve and Micah Lay, Bill Leonard, Bryce Maxell, Laura and Sebastian McMasters, Klaus Pirl, Ralph Powers, Lee Simons, Utah State University (Edmund Brodie Jr. and Jim McMahon), Janet Wagener, Woodland Park Zoo (Frank Slavins), and Neal Young.

Those who joined me in the field for turtle watching, merry lizard and snake chases, and long nights of road hunting were Eric Anderson, Jim Anderson, Rick Blair, Doug and Jeremiah Calvin, John Cossel Jr., Geoff Davis, Jeff Elsasser, Christian Engelstoft, Jim England, Bob Espinoza, Rick, Nathan, and Jasper Gerhardt, Brad Gommoll, Dave Hanson and Robin Dobson, Richard Hoyer, Brian Hubbs, Francis Jegou, Buck Jenkins, Doug Knutsen, Mike Labart, Steve and Micah Lay, Bill Leonard, Guy Lyons, Pete and Claudia Martin, Laura and Sebastian McMasters, Rich Nauman, Kristiina Ovaska, Chris Pearl, Chuck Peterson, Kelly Poetker, Tom and Casey Rodhouse, Mike Sequeira, Lee and Clayton Simons, Kate Slavins, Jan, Matthew, and Shawn St. John, Terry Steele, Bob Storm, Sue Summers, Karen Theodore, David and Jake Valverde, Jerome Verduyne, Janet Wagener, Gary and Ian Winter, and Craig Zuger.

Enormous appreciation goes to my family, who endured a long-term research project taking place in our home. Although they frequently were able to accompany me on field trips, more often they cheerfully wished me farewell and good luck as I routinely vanished into remote places in search of reptiles. My wife, Jan, is to be commended for reading the entire text and letting me know whether or not it made any sense.

Above all, I thank the Creator of our awesomely diverse natural world and all its creatures creeping and small.

Contents

Pictorial Guide

Turtles

Western Pond Turtle
p. 72

Painted Turtle
p. 76

Lizards

Zebra-tailed Lizard
p. 82

Great Basin Collared Lizard
p. 86

Long-nosed Leopard Lizard
p. 90

Desert Spiny Lizard
p. 94

Western Fence Lizard
p. 98

Eastern Fence Lizard
p. 102

Sagebrush Lizard
p. 106

Common Side-blotched Lizard
p. 110

Ornate Tree Lizard
p. 114

Desert Horned Lizard
p. 118

Pigmy Short-horned Lizard
p. 122

Greater Short-horned Lizard
p. 126

Western Skink
p. 130

Western Whiptail
p. 134

Southern Alligator Lizard
p. 138

Northern Alligator Lizard
p. 142

Rubber Boa
p. 148

Sharp-tailed Snake
p. 152

Ring-necked Snake
p. 156

Smooth Green Snake
p. 160

Racer
p. 162

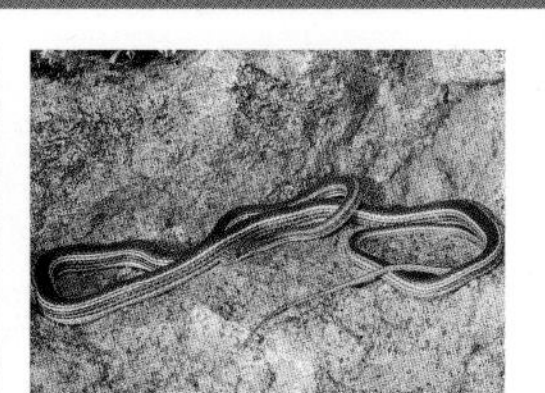

Striped Whipsnake
p. 166

California Whipsnake
p. 170

Coachwhip
p. 174

Western Patch-nosed Snake
p. 178

Pictorial Guide

Gopher Snake
p. 182

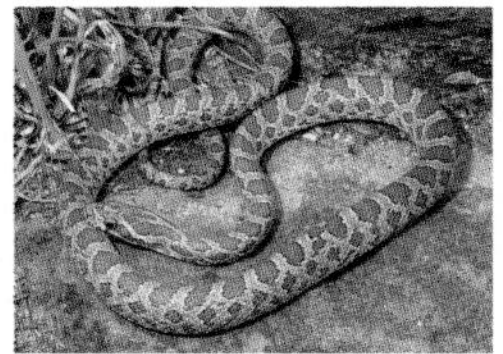

Corn Snake
p. 186

Common Kingsnake
p. 190

California Mountain Kingsnake
p. 194

Sonoran Mountain Kingsnake
p. 198

Milk Snake
p. 202

Long-nosed Snake
p. 206

Common Garter Snake
p. 210

Northwestern Garter Snake
p. 216

Western Terrestrial Garter Snake p. 220

Pacific Coast Aquatic Garter Snake p. 226

Sierra Garter Snake
p. 232

Ground Snake
p. 236

Night Snake
p. 240

Western Rattlesnake
p. 244

Foreword

Dr. Stebbins working on a painting for the latest edition of his field guide.

It was a distinct pleasure to be asked to write the foreword for this field guide. Many parts of this region are remote and poorly known to naturalists, especially in the northern Great Basin—these rugged wildlands offer a treasure trove for those willing to undergo the hardships that often accompany the excitement of new discoveries there. This, Al St. John has effectively done for its reptiles. Through his years of field work, insight, and photographic skill, much new information has come to light on the distribution, color variation, ecology, and other aspects of the lives of these fascinating animals.

I have read all his species accounts and have benefited from his gracious offer to extract items of interest for the revision of my own *Field Guide to Western Reptiles and Amphibians*. His accounts are well written and beautifully illustrated—the result, in large part, of his stature as a naturalist, his artistic sense, and his empathy with his subjects. Do not expect to find all the latest taxonomic revisions in this guide. He wisely withholds judgment, awaiting a period of peer review. A field guide is not the place for quick decisions on taxonomy, especially given the controversy over rapid-fire genetic-based revisions that are designating new species in the absence of adequate field studies.

A notable feature of his book is a section titled "field notes" that appears with each species account. These notes describe his on-site experiences in finding and observing the species covered in the book. As I read these stories, I felt I was with him in the field, experiencing the excitement of the hunt.

Field guides of broad coverage, such as my own and others, are no substitute for *regional* guides. The regional guide, narrower in scope, can provide far more detail and offer an intimacy with its subjects that books of broader coverage cannot hope to match.

I urge readers to acquire this guide and join Al St. John in expanding our understanding of the reptiles in a part of North America that contains some of our remaining truly great wild places.

Robert C. Stebbins
Emeritus Professor of Zoology and
Emeritus Curator of Herpetology,
Museum of Vertebrate Zoology,
University of California, Berkeley

Introduction

Painted Turtle Captive, Brad's World of Reptiles, Corvalis, Oregon.

Writing this field guide has been great fun because it has brought back childhood memories of searching for a special new book that would reveal to me the secrets of the world of reptiles. I remember capturing my first specimens at age five on my family's Willamette Valley farm in northwestern Oregon and the resulting questions these fascinating creatures caused me to ponder. There were three snakes with various colored stripes and a large, foul-tempered lizard with a long tail. I put them all in an open-topped cardboard box equipped with dirt, grass, and a bowl of water to make them feel at home. I squatted beside the make-shift enclosure for a considerable time and watched the reptiles with rapt attention. What were their names? I hadn't a clue, nor could my parents or grandparents tell me for sure. Certainty was lacking and I had a burning desire to identify them and learn about their life histories.

A shelf of herpetological classics.

On my sixth birthday, enlightenment arrived. My parents gave me the gift of a paperback book titled *Reptiles and Amphibians: A Guide to Familiar American Species.* Fresh off the presses, this 1953 book by Hobart Smith and Herbert Zim was the first edition of a wonderful little publication in the Golden Nature Guide series. There were beautiful color illustrations by James Gordon Irving, along with a short text description and distribution map for each kind of amphibian and reptile. To this day, I distinctly recall how I tore off the wrapping paper, ran outside, sat on the lawn in the warm autumn sunshine, and eagerly looked at every page. Mom helped me with the reading later, and I learned that my former reptilian captives were a species of garter snake and a Western Alligator Lizard (presently called the Southern Alligator Lizard.) Before long, I had learned what I could from the book and wanted to know more. My resulting quest undoubtedly mirrors the experiences of many other fledgling naturalists during the 1950s and 1960s and somewhat traces the history of publications that influenced field herpetology in western North America.

My next herpetological boon came at around 12 years of age when I was thrilled to discover that I could check out books from the state library in Salem by mail. Soon, the unfortunate postman was periodically lugging heavy boxes to our door that contained imposing, two-volume sets of such classics as

The Reptiles of Western North America, by John Van Denburgh (1922); *Rattlesnakes: Their Habits, Life Histories, and Influence on Mankind*, by Laurence Klauber (1956); and *Handbook of Snakes of the United States and Canada*, by Albert Wright and Anna Wright (1957). Of particular interest to me was a 1939 Oregon State College monograph by Kenneth Gordon, *The Amphibia and Reptilia of Oregon*. I mined that book for all it was worth, learning more precisely where each species and subspecies ranged in my home state and the habitats they preferred. Complimenting Dr. Gordon's publication was Robert Stebbin's landmark reference book, *Amphibians and Reptiles of Western North America* (1954).

In 1966, the year after I graduated from high school, Dr. Stebbins produced what became the bible of field herpetologists in the American West, *A Field Guide to Western Reptiles and Amphibians*. This addition to the excellent *Peterson Field Guide Series* was jam-packed with a wealth of useful information and illustrated with the author's superb paintings and pen-and-ink drawings. Compact and ruggedly hardbound, that prized book went everywhere with me. At last, my gnawing hunger for knowledge had been satisfied by this succession of top-quality literature.

I certainly do not have the audacity to think that my book is in the same league as the preceding scholarly publications. I am hopeful, however, that it will be a useful contribution to our understanding of the region's reptiles. Despite my overall intention to produce a field guide for both children and adults, I couldn't resist repeatedly asking myself the question, "What would a 12- to 15-year-old kid obsessed with the desire to know more, like I was at those ages, want me to share?"

Obviously, providing fast and accurate identifications of reptiles should be the primary purpose of a field guide. However, if someone doesn't have a clue where and how to find these creatures, it will be a less useful publication. I think a good field guide should be just that, a guide that actually instructs how to find each species in the field. With that in mind, I have included in every species account as much information about habitat preferences, distribution, and behavior as is possible within the confines of this compact format. More in-depth information on these subjects is given in the "Reptile Habitats in the Northwest" and "Observing Reptiles in the Field" sections. Those two parts of the book have more lengthy texts and are best studied prior to going reptile-watching. Preparing beforehand with pertinent knowledge can often make the difference between success and failure. Additionally, each species account is accompanied by a "Field Notes" section that provides a short anecdote from my personal experience with that particular reptile species. These stories convey something of the "flavor" of being out in the field and give helpful hints on techniques for finding and studying reptiles.

Children are fascinated by reptiles, especially snakes. Matthew St. John and a Common Kingsnake

Introduction

I realize that my field notes stray from the usual format of field guides published during recent decades, but from the late 19th century through the 1950s, personal anecdotes were commonly used in many natural history books. I remember that to my 12-year-old mind, the best parts of Wright and Wright's *Handbook of Snakes* were their field notes sections included for nearly every species, and I'm certain that I wasn't the only youngster whose imagination was electrified by reading the following account about the California Mountain Kingsnake in Raymond Ditmars' *Reptiles of North America* (1907): "While collecting in the San Bernardino Mountains, I found several of these snakes under the loose bark of big, fallen pines. To strip away a slab of bark and discover one of these lustrous, handsome creatures is to feel that one has a real prize." Those two lively sentences teach so much in such an entertaining way. However, the writings that probably galvanized a whole generation of budding herpetologists toward careers in field research like no other were Carl Kauffeld's autobiographical stories in *Snakes and Snake Hunting* (1957). Nearly every herpetologist that I have talked with who grew up during that era told me that they, too, were totally enthralled by his adventures and nearly ran away from home to become vagabond snake hunters in the deserts and mountains.

While a preserved museum specimen certainly has its value, there is nothing quite like observing an animal in its own unique habitat going about the business of daily survival. Nature study needn't be merely the dry, academic accumulation of scientific data. It can be undertaken in a zestful, fun-filled manner. Children are especially enthusiastic about reptiles. A family outing is a wonderful chance to learn how to use a field guide. This skill will provide the key to a lifetime of learning about the creatures that share the world around us, whether it be in a remote wilderness or in our backyards.

I should add here that despite my dedication to the neophyte by not bogging things down in arcane technicalities, I have also provided a considerable amount of information that should be useful to professional herpetologists, biologists, and other scientists. It is my sincere hope that I have been successful in meeting the needs of both groups.

This field guide represents more than 30 years of herpetological study. Initially, much of this experience was gained in Oregon; later, I began to roam beyond the borders of my state and explore other areas in the Northwest. Signing a contract with Lone Pine Publishing in 1997 marked the beginning of a particularly intense four-year period of field work. I had to find and photograph all the turtles, lizards, and snakes included here, and I made numerous trips throughout the region. Sometimes the search for an especially elusive species took weeks, months, or even years. I enjoy this sort of challenge; like all naturalists, I'm happiest when I can spend a lot of time outdoors. I'm made that way. So is this book. It's durable and field-oriented in its design. Please toss it in your pack when you head out the door for a hike. Happy trails!

Great Basin Collared Lizard
Pyramid Lake, northwestern Nevada.

How to Use This Field Guide

The area covered by this book is defined by the natural features of the Northwest rather than political demarcations. I have used the Continental Divide as the eastern boundary of the region (except in northern British Columbia, where I followed the spine of the Rockies) and the Pacific Ocean as the western. The northern limit is at roughly the 58th parallel, where it becomes too cold for reptiles. The southern boundary of the Northwest is more difficult to define, because varying elevations and climates create a complex, ragged border. I chose to use the 40th parallel because most Northwest plant communities fade away in that vicinity. In the United States, this field guide covers Washington, Oregon, Idaho, western Montana, southwestern Wyoming, northwestern Colorado, the northern parts of Utah, Nevada, and California, and extreme southeastern Alaska. In Canada, most of British Columbia is included.

Within this vast region, 42 species of native reptiles have been recorded. This assemblage comprises 2 turtles, 16 lizards, and 24 snakes. There are also three species of marine turtles native to the Pacific Ocean, but I have not included them because they are so rarely seen along our shores. Additionally, there is a section at the end of the species accounts that covers three introduced reptiles that have established breeding populations in the Northwest. Combining both native and introduced reptiles, there are accounts for 45 species.

The belly scale pattern on a Western Fence Lizard.

The shed skin of a Western Terrestrial Garter Snake. Wasatch Range, northwestern Utah.

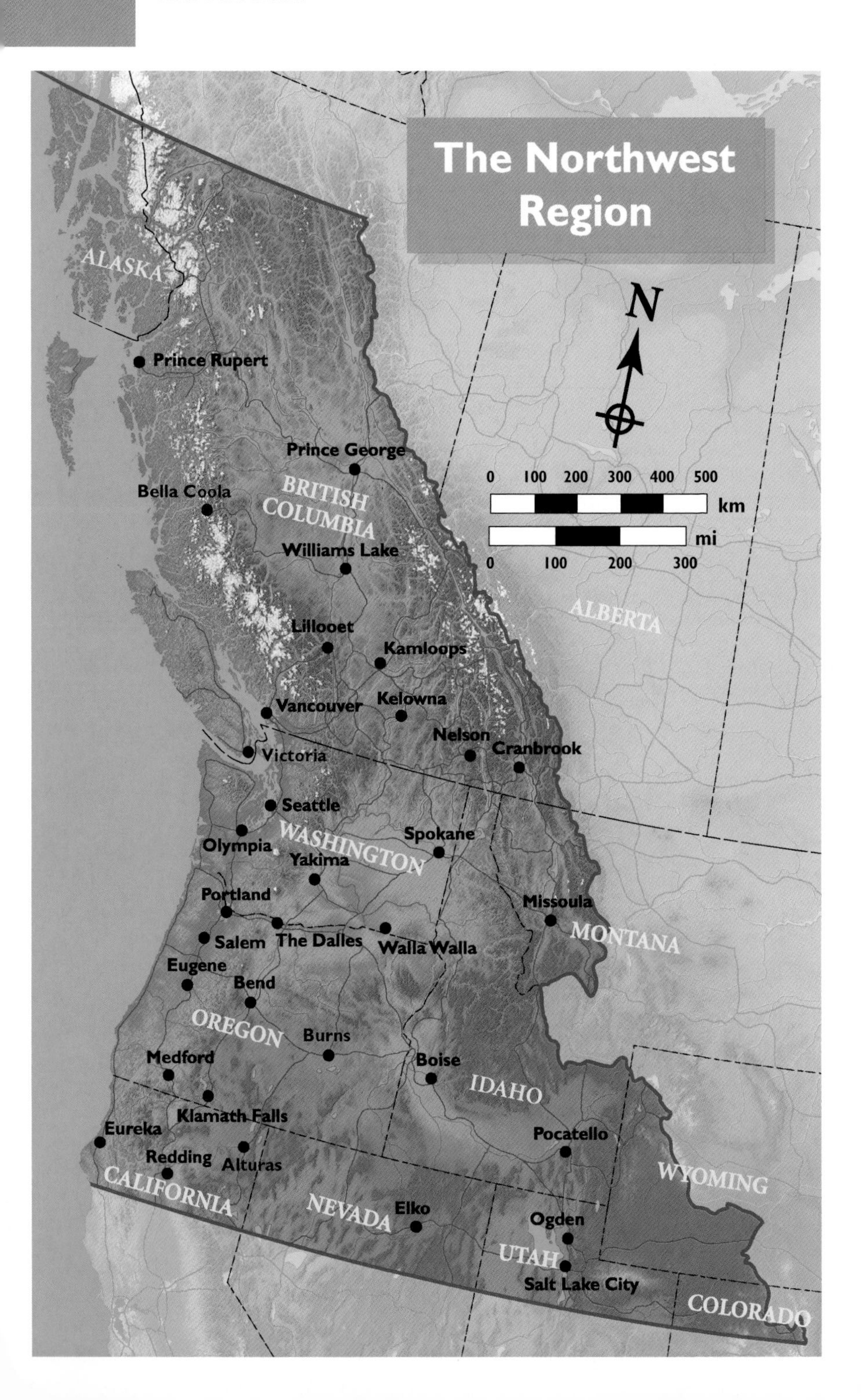
The Northwest Region
N
0 100 200 300 400 500
km
mi
0 100 200 300
ALASKA
Prince Rupert
Prince George
Bella Coola
BRITISH COLUMBIA
Williams Lake
Lillooet
Kamloops
Vancouver
Kelowna
Nelson
Cranbrook
Victoria
ALBERTA
Seattle
Olympia
WASHINGTON
Spokane
Yakima
Portland
Missoula
Salem
The Dalles
Walla Walla
MONTANA
Eugene
Bend
OREGON
Burns
Medford
Boise
IDAHO
Klamath Falls
Eureka
Pocatello
Redding
Alturas
CALIFORNIA
WYOMING
NEVADA
Elko
Ogden
UTAH
Salt Lake City
COLORADO

A Western Fence Lizard living up to its name. Near Rowena, Columbia River Gorge National Scenic Area, Oregon.

This field guide provides morphological characteristics for the identification of species, along with information about distributions, habitat preferences, behavior, and suggested methods of field study. Under the following headings, I explain how to use the various parts of the keys and species accounts.

IDENTIFICATION KEYS

Immediately after this introduction there are "Quick Keys" to the three reptile groups: turtles, lizards, and snakes. These identification aids are primarily visual, but brief descriptions accompany the photos to aid in identification. Small drawings of pertinent characteristics may be included for further clarification.

As an example, if the reptile you are trying to identify is a snake, turn to the first page of the Snake Quick Key section (p. 63). The "Quick Key Finder" you will see there pairs brief descriptions with iconic drawings that depict four basic patterns of snake coloration: uniformly colored, blotched, crossbanded, and striped. Let's say the snake in question has vivid black and white encircling crossbands. Because the crossbanded icon most closely resembles the pattern of your snake, you would choose that as the primary identifying characteristic. Next to the crossbanded snake icon will be a page number. When you turn to that page, you will see small color photos of all the crossbanded snakes found in the Northwest, arranged for easy comparison. You decide that the photo of the Common Kingsnake (*Lampropeltis getula*) most closely resembles your specimen. Next to that photo will be the page number of its species account, where a more detailed description is given along with additional photos, a distribution map, and a full life history. The Quick Keys for turtles and lizards work in the same manner.

The Quick Keys include photos of most subspecies (unless they are very similar in appearance) and other significant variations of a species known to occur in the Northwest to provide as thorough a comparison as possible. Most average reptiles can usually be rapidly "keyed out," but you will occasionally find an individual that does not fit the norm, particularly in the case of garter snakes. In such situations, choose the species that most closely resembles your specimen and turn to its account for more detailed information. Once you become familiar with how to use the Quick Keys, identifying reptiles will be fun and easy.

Introduction

SPECIES ACCOUNTS

Names

I have usually followed the recommendations given in *Scientific and Standard English Names of Amphibians and Reptiles of North America North of Mexico* (Crother et al. 2000). Not all herpetologists are in total agreement on many of these choices and a number of recent taxonomic changes, however, and two names that are in popular usage may be given for some species. As much as possible, I have aligned this guide with the common and scientific names used in other currently published books about reptiles.

Identification

To fit the format of a compact field guide, only the primary distinguishing characteristics for each species are given. From my experience, most people will use this book while grasping a wriggling reptile on a steep, rocky slope or precariously balanced on a log in a wetland. The emphasis here is to give only the pertinent facts to promote speedy identification. The most important characteristics have been placed in **boldface** type, and I have tried to keep the herpetological jargon to a minimum. Only a few useful technical terms have been incorporated, such as "dorsal" and "ventral." A glossary of these words is provided on page 253. In addition, drawings that illustrate the various anatomical parts of reptiles are provided on pages 56–69. Adult sizes of turtles refer to the overall length of the shell, while snakes are measured from the tip of the nose to the tip of the tail. Because many species of lizards easily lose their tails, two adult lengths are given: from the tip of the nose to the vent (snout-vent length), and from the tip of the nose to the tip of the tail (total length). Complicated methods for identification, such as body-scale row counts on snakes, have not been used. These practices are better suited to the laboratory with the aid of a magnifying glass. Out of necessity, simple counts of upper labials (lip scales) on garter snakes have been used to sort out our five species within this confusing group. These large scales are generally easy to count. If male and female individuals can be readily distinguished, characteristics for differentiation are given. Likewise, when juveniles differ greatly in coloration and pattern, a description is included.

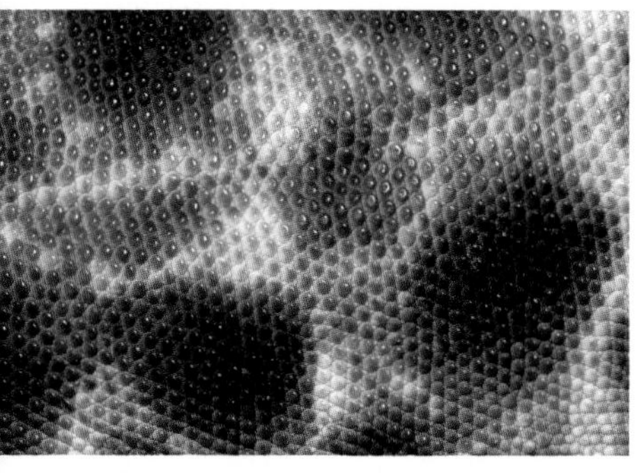

The dorsal pattern on a Long-nosed Leopard Lizard.

Variation

Many species of reptiles vary in appearance throughout their geographic distributions. Morphological traits sometimes differ so greatly that an observer could think that two dissimilar individuals are from entirely different species. Generally, when a certain variation is uniformly seen throughout all populations within a part of the species' range, it may be named as a subspecies (or race). Where the ranges of two subspecies meet, individuals with characteristics of both will usually be encountered. These zones of intergradation may be relatively narrow or sometimes quite wide, occasionally spreading over a region of 100 mi (160 km) or more. These broader areas of intergradation are mentioned in the text and illustrated on the accompanying distribution maps. I have also used this section to describe morphological variations of individuals within single populations, which are usually referred to as "morphs." The Northwestern Garter Snake (p. 216) is a good example, with its array of different colored stripes and patterns. Often, one litter of young may contain a varying collection of individuals with red stripes, yellow stripes, or turquoise stripes. Any other significant geographical variations within a species (such as widespread melanism in certain populations) are covered here as well.

Over the years, subspecific classifications are often changed as the result of new information gained through research. One subspecies may be deemed invalid, while another may be elevated to full species status. These taxonomic changes have become more frequent in recent years with the use of molecular techniques that reveal the genetic relationships among groups of lifeforms. Genetic findings sometimes seem to support past classifications of species and subspecies that were based solely on morphological

An autumn rattlesnake hunt in the Willamette Valley, northwestern Oregon.

characteristics; in other instances, two reptiles that look identical but are from different regions may turn out to be entirely different genetically.

The science of naming and classifying organisms through taxonomy and systematics is in a state of flux. Opinions differ. Many scientists question the old concepts of what actually constitutes a species or subspecies, and some recent taxonomic changes have led to disagreements and debates. I have tended to be conservative with some of the newest designations (or eliminations) of species and subspecies, along with transfers to new genera, so the names and descriptions listed in this book mostly reflect longer-held divisions or changes that have been unanimously accepted by herpetologists. I indicate where major taxonomic disagreements have occurred, and I briefly explain the differing points of view. If a longer period of peer review reveals a general acceptance of recent findings, I plan to update the names in future editions of this book.

Similar Species

Although no two species of reptiles are totally identical in appearance, some closely resemble each other in various ways that could cause confusion. As an aid in sorting out these look-alikes, this section briefly describes the clarifying diagnostic characteristics so they can be easily differentiated. Again, as in the Identification section, the text has been streamlined to include only the pertinent, needed facts.

Distribution and Maps

I have limited my written description of a species' distribution to the Northwest. Along with the primary portion of each reptile's range, I also note disjunct populations, isolated single records, anomalous past locality records that are in need of updated clarifications, unconfirmed reports, and elevational distributions where appropriate. The main map provided for each species illustrates that reptile's distribution in the Northwest. All recognized subspecies that occur in our region are designated with different colors (also numbered as an aid to people who have difficulty distinguishing colors). Significantly broad areas of intergradation between subspecies are designated by alternating stripes of the two subspecies' colors. Isolated single records are marked by black dots. Unconfirmed reports are indicated with question marks. A smaller inset map shows the species' overall range in North America.

When using the maps, you should keep in mind that the range of a species is never as continuous as the colored shadings imply. In reality, reptile distributions are usually spotty, with numerous interlacing areas of unsuitable habitat. As an example, the Desert Horned Lizard's map (p. 119) gives the impression that this species is uniformly distributed throughout the northern Great Basin. This reptile only inhabits arid, sandy-soiled basins and foothills, however, and it is absent from the higher, rocky ranges that intervene throughout the area. When populations of a reptile species occur at locations widely separated from the primary distribution, they are shown as disjunct ranges on the map.

Habitat and Behavior

The intention of this section is to provide the information you need to go out into the field, locate the appropriate habitat for the reptile you are interested in seeing, and look for it in the right places and at the right time of day. Details of behavior (including feeding and reproduction) and specific micro-habitat needs are given to help achieve this goal. You will learn if the reptile is active during the day (diurnal) or at night (nocturnal), and whether it hides under objects or will be out sunning on rocks and logs. When specific soil types, rocks, or plants are closely associated with a certain species of reptile, I have listed them. Because plants are mentioned so frequently in various parts of this field guide, only their common names are used to avoid cluttering the text. A list of all the mentioned plants with their scientific names can be found on page 252. Further information about the environments inhabited by turtles, lizards and snakes, along with field techniques for studying them, are provided in the sections titled "Reptile Habitats in the Northwest" (p. 27) and "Observing Reptiles in the Field" (p. 42).

Field Notes

At the end of each species account I include a short anecdote from my personal experience with that kind of reptile. I hope these stories are somewhat entertaining, and that they give you some useful tips on where, when, and how to find and study that species. Often, these anecdotes impart something of what's unique about a particular reptile, as well as provide a better understanding of what a field herpetologist does. They also suggest some good places in the Northwest where various species of reptiles can be seen. It is an added bonus that many of these places are located in some of the most beautiful scenery on the planet. Generally, I have mentioned areas that are under protection and open to the public, such as government wildlife preserves or areas managed by a private conservation group like The Nature Conservancy. In many instances, however, I have been vague about the exact location when the story takes place on private land or in the fragile habitat of a rare species.

A Western Rattlesnake in a defensive pose. Near Hilt, Klamath River drainage, northern California.

About Reptiles

A field-oriented guide such as this book is not the place for an in-depth overview about reptiles. There are many excellent books available that provide a more comprehensive treatise for anyone wishing to learn what is unique about reptiles compared to the other animal groups. Some of these publications are listed in the "Resources" section. To provide some easily accessible, general background information about reptiles, however, the following section answers some of the most common queries that people have about reptiles.

Oregon Spotted Frog (Rana pretiosa). *Davis Lake, Cascade Mountains, Oregon.*

Is it a reptile or an amphibian?

Many people confuse reptiles with amphibians. In the Northwest, reptiles are represented by turtles, lizards, and snakes, and amphibians by salamanders, toads, and frogs. Misunderstandings are particularly common concerning salamanders, which are often thought to be some type of lizard. Reptiles have dry, scaly, or plated skin; amphibians are usually moist-skinned (although toads and newts have relatively dry skin), don't have any scales or plates, and lack claws on the toes. Generally speaking, amphibians prefer cool, damp, or wet habitats, whereas reptiles are more often found in warm, dry areas, although in our region, turtles and some garter snakes are largely aquatic. However, even they spend a good portion of their time sunning on dry logs, rocks, or banks.

Ensatina Salamander (Ensatina eschscholtzii). *Chetco River drainage, southern Oregon coast.*

Western Toad (Bufo boreas). *Cascade Mountains, Oregon.*

A pair of Great Basin Collared Lizards elevating their body temperatures in the morning sunshine. Pyramid Lake, northwestern Nevada.

Night Snakes are nocturnal hunters that can be active during cooler conditions than most reptiles can tolerate. Malheur National Wildlife Refuge, southeastern Oregon.

A Zebra-tailed Lizard curling its toes and tail upward from hot sand. Pyramid Lake, northwestern Nevada.

Why are reptiles called "cold-blooded" animals?

Reptiles, which are properly called "ectothermic" animals, cannot, to any large degree, internally regulate their body temperature, as can mammals and birds, which are "endothermic." Reptiles are usually much the same temperature as their surroundings, and they control their body temperatures through behavior, moving about between sun and shade to avoid extremes of hot and cold. On a cold day a reptile will be cold and lethargic. On a warm day it will be warm and lively. When the intense heat of summer prevails, many snakes become primarily active at night. During the winter months in the Northwest, reptiles enter an inactive state of hibernation in various types of underground retreats. Conversely, if prolonged hot, dry periods occur during the warm season, some reptiles retreat into a summer hibernation called "estivation." Because reptiles rely on solar heat for warmth, they do not need to be constantly supplying themselves with food for calories to produce metabolic heat. Unlike birds and mammals, reptiles can survive on amazingly small amounts of nourishment, which also allows them to inhabit extremely arid, barren areas with little available food.

A Western Rattlesnake using its heat-sensory pits (the round openings on each side of the head). The forked tongue collects scent particles.

What senses do reptiles use?

Although most lizards have relatively good hearing, snakes are not as well equipped. Snakes lack outer ear openings and eardrums, and they have only rudimentary inner ears. On the other hand, snakes are very sensitive to ground vibrations. It was long thought that snakes were completely deaf, but studies have revealed that they do hear low-frequency, airborne sounds transmitted to the inner ear by certain head bones. Consequently, a rattlesnake probably cannot hear its own rattle, because it is a higher-frequency sound.

As for vision, most lizards have good eyesight, whereas many species of snakes have limited vision, being rather near-sighted. There are exceptions among various kinds of snakes that have relatively better distant eyesight, such as members of the racer and whipsnake groups (genera *Coluber* and *Masticophis*), which are diurnal, visually oriented predators. However, studies indicate that they cannot clearly see stationary objects and require movement to detect prey or enemies. Turtles have excellent vision, but poor hearing.

Most reptiles have an acute sense of smell. The snakes of the pit viper family, of which rattlesnakes are members, have another interesting detection system. They are equipped with small heat-sensory pits on each side of their head between the eye and the nostril. These pits are extremely sensitive to infrared radiation and aid in making an accurate strike at warm-blooded prey, even in the dark. Many species of lizards have what is sometimes referred to as a "third eye." This parapineal organ is a light-sensitive, transparent disk on top of the head. It is connected by nerves to the brain through an aperture in the skull, and research indicates that it probably helps regulate the amount of time a lizard spends in the sun.

Does a snake sting with its tongue?

No. A snake's tongue is not the poisonous, defensive device that many people believe it to be. It is harmlessly used for sensing a snake's surroundings. When a snake sticks out its forked tongue it picks up scent particles that are pulled back into the mouth on the surface of the tongue and deposited on two small indentations in the roof of the mouth. These cavities are the vomeronasal organs (sometimes called the Jacobson's organs) and are very sensitive, analyzing the scent particles and sending signals to the olfactory part of the snake's brain.

What and how do reptiles eat?

The majority of reptiles are carnivorous, eating small animals that range in size from insects and slugs to frogs and young rabbits. The choice of food varies greatly from one species to another and depends on the size and age of the individual reptile. Turtles and certain species of lizards also include plants in their diet. Some snakes are called "cannibalistic" because they often eat other snakes. In the Northwest, the Common Kingsnake is most noted for this habit. Because this beautiful black-and-white, crossbanded serpent is immune to the venom of rattlesnakes, its menu occasionally includes these poisonous reptiles. Many lizards eat other lizards, often of their own species.

The Common Kingsnake sometimes climbs trees in search of the eggs and nestlings of birds. Near Hilt, Klamath River drainage, northern California.

Most reptiles have pointed teeth of some type for biting and holding food. Turtles lack teeth, having instead a sharp, parrot-like beak. Lizards and turtles often chew their food to some extent before swallowing it, whereas snakes swallow their food whole. The fact that snakes can swallow food items that are often much larger than the size of their head never ceases to cause amazement.

A Long-nosed Leopard Lizard swallowing a juvenile of its own species; only the tail remains. Alvord Basin, Oregon.

A Great Basin Collared Lizard stalking and capturing a grasshopper. Pyramid Lake, northwestern Nevada.

This ability is possible because a snake's jaws are only loosely attached to the skull by stretchy tissue and a long, vertical bone at the rear of each lower jaw. Also, an elastic ligament connects the ends of the jaw at the chin. This flexible combination makes for an expanding hinge that can stretch both vertically and horizontally, allowing the mouth to open to the maximum.

Reptiles employ various methods to capture their prey. Most species simply grasp the intended food in their jaws and swallow it alive. One of the most interesting methods is used by the constricting snakes. Two or more loops of the snake's body are wrapped around an animal. Then, squeezing tightly, the snake compresses the victim's chest cavity until the lungs are deflated and the heart prevented from beating, leading

A Gopher Snake constricting a rat. Captive reptile, High Desert Museum, Bend, Oregon.

quickly to death. Venomous pit vipers, such as rattlesnakes, have two enlarged, hollow fangs connected to venom glands that function like hypodermic syringes and needles. They simply bite their prey and wait for it to become paralyzed or die before eating it. Rear-fanged snakes, such as the Night Snake, have enlarged, grooved teeth at the back of each jaw. When the snake chews on a small animal intended for food, venom runs down the grooves into the wound (harmless to humans.) Some species of garter snakes, and possibly the Ring-necked Snake and Ground Snake, have mildly toxic saliva to help subdue struggling prey. Because they can accommodate such large meals at one feeding and have low metabolic needs, snakes do not have to eat often. In the wild, snakes may eat only once a week or less. In captivity, snakes have been known to go well over a year without food, living off their stored body reserves.

A Gopher Snake eating a rat. Captive reptile, High Desert Museum, Bend, Oregon.

How do reptiles shed their skins?

Reptiles have an interesting way of discarding their old, dead outer layers of scaly skin, or in the case of turtles, the individual shields of the shell. As new scales form under the surface layer of the skin of a snake or lizard, there is a loosening action that causes the older skin to be cast off. Most lizards shed their skin in strips, much the way we do after a sunburn. Snakes shed their skin in one entire piece, including a protective covering of clear scales over the eyes. Because snakes

A Northwestern Garter Snake with hazy blue eyes prior to shedding. Near Corvallis, Willamette Valley, northwestern Oregon.

A large section of shed Northern Alligator Lizard skin. Howard Prairie, southern Cascade Mountains, Oregon.

do not have eyelids, these "eye caps" (spectacles) protect the eye and fit tightly over it like contact lenses. It is very important that these old caps come off, otherwise they will become opaque, dead skin that might eventually blind the snake. To insure that the eye caps slide off with the skin, the snake secretes a lubricating fluid under the skin and eye caps just a few days before shedding. This fluid causes the eyes to become a hazy blue and dims the snake's vision. When it is ready to shed, the snake rubs its snout on rocks, branches and other objects in its environment, breaking the

A Common Kingsnake shedding its skin. Captive reptile.

skin loose along the lips and working it off the head; then it crawls out of the old skin. The skin usually comes off inside-out, in much the same manner as we might pull off our socks. Most reptiles shed their skin several times each year, depending on their rate of growth.

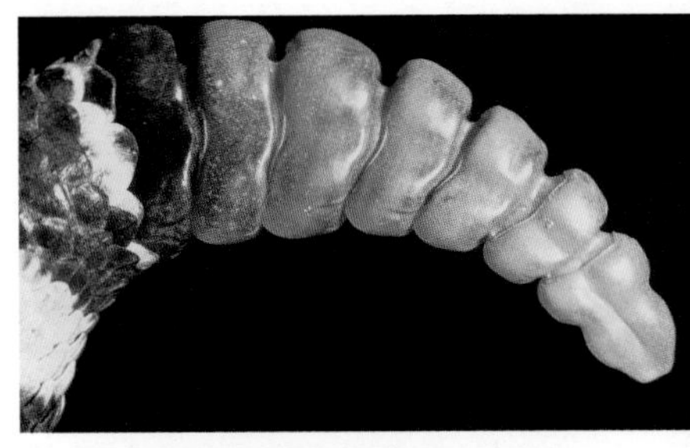

The rattle of a Western Rattlesnake.

Can a rattlesnake's age be computed by counting its rattles?

No. A new segment is added to the base of the rattle each time the skin is shed, but a rattlesnake may cast its old skin more than once a year. Also, when the rattle becomes longer and more fragile, several segments may break off at the end and be missing.

Are snakes slimy?

No. All snakes have dry, clean skins. Their bodies are rather pleasant to touch, feeling much like high-quality leather.

Snake skin has a pleasant feel, much like leather. Shawn St. John with a Gopher Snake, Columbia River Gorge National Scenic Area, Oregon.

Great Basin Collared Lizards mating. Pueblo Mountains, southeastern Oregon.

Eggs of a Western Pond Turtle. Willamette Valley, northwestern Oregon.

How do reptiles reproduce?

In the cooler, northerly latitudes of the Northwest, reptiles usually mate during the spring, or in early summer at high elevations. Males often seek females immediately after emerging from hibernation, when they are near the communal denning sites. Many male lizards perform courtship displays of head-bobbing and "push-ups." Commonly, male lizards also develop seasonal breeding colors, but there are several species in which the females become more colorful. Male snakes use their sense of smell to track the pheromone trails of females. These powerful secretions stimulate a strong sexual attraction, sometimes causing several male snakes to simultaneously attempt to mate with a single female. Male Western Rattlesnakes will occasionally perform a "combat dance," in which they rear up, entwine, and attempt to wrestle each other to the ground. In a number of species of reptiles, the sperm is sometimes stored in the female's oviduct after mating until the following spring. There are some turtles and snakes that can store viable sperm for several years.

A Common Kingsnake hatching. Captive reptile, Brad's World of Reptiles, Corvallis, Oregon.

Although most reptiles are "oviparous" (they lay leathery-shelled eggs), many species of lizards and snakes are "viviparous" (they bear live young). Reptiles deposit their eggs in rotting logs, soil, sand, or burrows where the heat of the sun or decomposing plant material can incubate them. Turtles leave the water and bury their eggs in a hole they dig on a sunny southern exposure. (As an interesting side note, the temperature of turtle eggs during incubation determines the gender of the developing turtles). Depending upon the species and various environmental influences, hatching or birth usually does not

A female Western Skink brooding eggs. Chetco River drainage, southern Oregon coast.

occur until later in the summer or early fall. Most reptiles do not remain with their eggs or offspring, but abandon them immediately—the young are ready to fend for themselves shortly after entering the world. The exception among this region's reptiles is the female Western Skink, which will remain with its eggs until they hatch. An introduced lizard found in central Oregon, the Plateau Striped Whiptail, is an all-female species that reproduces by parthenogenesis (the ovum develops without fertilization by a male.)

Jim Riggs examining a dead Western Rattlesnake that was run over by an automobile. Catlow Valley, southeastern Oregon.

What are their enemies and how do they defend themselves?

The main enemies of reptiles are predatory mammals, birds, and other reptiles. Humans, however, who often misunderstand, fear, and kill reptiles, pose one of their greatest dangers. Another major threat is dying under the wheels of an automobile on our ever-spreading maze of roads. Poisonous sprays and herbicides also take their toll. Although most reptiles will bite when attacked or caught (the rattlesnake, with its venom-delivering apparatus, being the most effective), all will usually try to first avoid their enemies through a quick escape. Other evasion tactics depend upon the species and include hiding through protective camouflage coloration that blends with the surroundings, playing dead until the danger is gone, employing a tail that distracts as a decoy in some way, and in the case of certain snakes, rolling into a ball and hiding their head under the protective coils. Some snakes attempt to intimidate a perceived enemy by flattening the head, hissing, and striking. This type of defensive display is often accompanied by vibrating the tail, the rattlesnake being particularly notable because of the startling "buzz" it produces. Most species of lizards have the additional advantage that their tails easily detach when grasped, allowing them to escape. Unlike snakes, lizards have the ability to regenerate a new tail. Many lizards can change color or lighten and darken their skin, which can aid

The stripes on this Common Garter Snake provide protective camouflage in vegetation. Umpqua Valley, southwestern Oregon.

A Common Side-blotched Lizard with a missing tail. Owyhee River Canyon, south-eastern Oregon.

in avoiding detection by enemies. Both lizards and snakes will sometimes discharge foul-smelling feces to discourage a captor, and snakes create a particularly repellent mixture by adding a strongly repugnant musk that is emitted from scent glands in the cloaca. In many species, this musk also causes irritation and pain when it comes into contact with a predator's eyes. Turtles often release urine from their bladder when picked up.

Are venomous coral snakes, Copperheads, Cottonmouth Water Moccasins, and Gila Monsters found in our region?

No. There is only one dangerously venomous reptile native to the Northwest, the Western Rattlesnake.

Some harmless species of snakes, such as this Milk Snake, resemble a venomous coral snake. Stansbury Mountains, northwestern Utah.

First Aid For A Rattlesnake Bite

Controversy surrounds the issue of what is the best immediate treatment for a rattlesnake bite. The long-accepted methods of using a tourniquet, making small incisions through each fang puncture, applying suction to the wound, and, finally, using cold packs (cryotherapy) has fallen into disfavor in recent years. The misuse of these methods by the untrained has resulted in numerous problems.

Most poison control centers now give the following advice to the general public:

- Keep the area of the bite close to the level of the heart (i.e., if bitten on the hand, restrain the arm in a sling next to the chest; if bitten on the ankle, lie down).
- Remain as immobile as possible (splinting a bitten arm or leg is an effective way to arrest movement entirely). If feasible, bring medical aid to the patient.
- Try to keep the victim calm and reassured; the bite of the Western Rattlesnake is rarely fatal to healthy adults. (Children are more susceptible because of their smaller size.)
- Contact emergency help as soon as possible.

There are excellent poison control centers in the major cities of the Northwest. They maintain contact with ambulance and helicopter rescue services, community hospitals, local and national specialists in all areas of poison management, and where antivenin may be obtained throughout the region. They have 24-hour hotline numbers. Check the inside front cover of a phone directory or call 911.

Reptile Habitats in the Northwest

The Cascade Mountains are the major range that divides the Northwest into two completely different climates. Scout Lake, Mt. Jefferson Wilderness, Oregon.

Despite the northerly latitudes of the Northwest, there is a fairly good representation of reptile species native to this vast section of the North American continent. The bountiful ecological diversity created by the region's numerous mountain ranges and their resultant climatic modifications are the primary cause. Most of these ranges run north and south, with valleys, basins, canyons, and rolling plateaus and plains situated between. Although the rain-drenched, foggy coastal areas have a limited number of reptile species, sufficiently sunny conditions for more of these animals occur throughout most of the interior areas. Moving inland to the east of the mountain ranges, with their subsequent "rainshadows," progressively drier habitats are encountered. Generally, west of Oregon and Washington's Cascade Mountains, British Columbia's Coast Mountains, and California's northern Sierra Nevada, the winters are relatively mild, and the summers are frequently moderated by cooling winds from the Pacific Ocean. To the east of these mountains, in the high, interior plateau areas, there is a continental climate of cold, snowy winters and hot, dry summers. A herpetologist who travels north or south in the Northwest will also encounter differences. The cooler climate of Canada sustains only a few kinds of turtles, lizards, and snakes. Conversely, the warmer latitudes near the southern edge of our region have a larger reptilian variety and harbor the northernmost limits of many species typical of the reptile-rich American Southwest. Additionally, elevational gain limits reptile life, with few species ranging into the cold, harsh zones of mountains above 6,000 ft (1,830 m).

Because the Cascades, Sierra, and British Columbian Coast Mountains divide the Northwest into two distinctly different climates, there is a noticeable separation of species. Reptiles adapted to damper coniferous forests and oak woodland habitats occur primarily west of these mountains; those species that prefer more arid environments inhabit the drier pine-juniper woodlands and open sagebrush steppe or salt scrub desert communities found to the east. There are several exceptions of species that range widely throughout large sections of the regions, on both sides of the mountains. In adapting to these vastly different environments, however, some of the reptiles are recognized to have diverged into distinct subspecies that are morphologically different in various ways and are unique to the eastern or western sides of these climatic divides.

Ecoprovinces of the Northwest

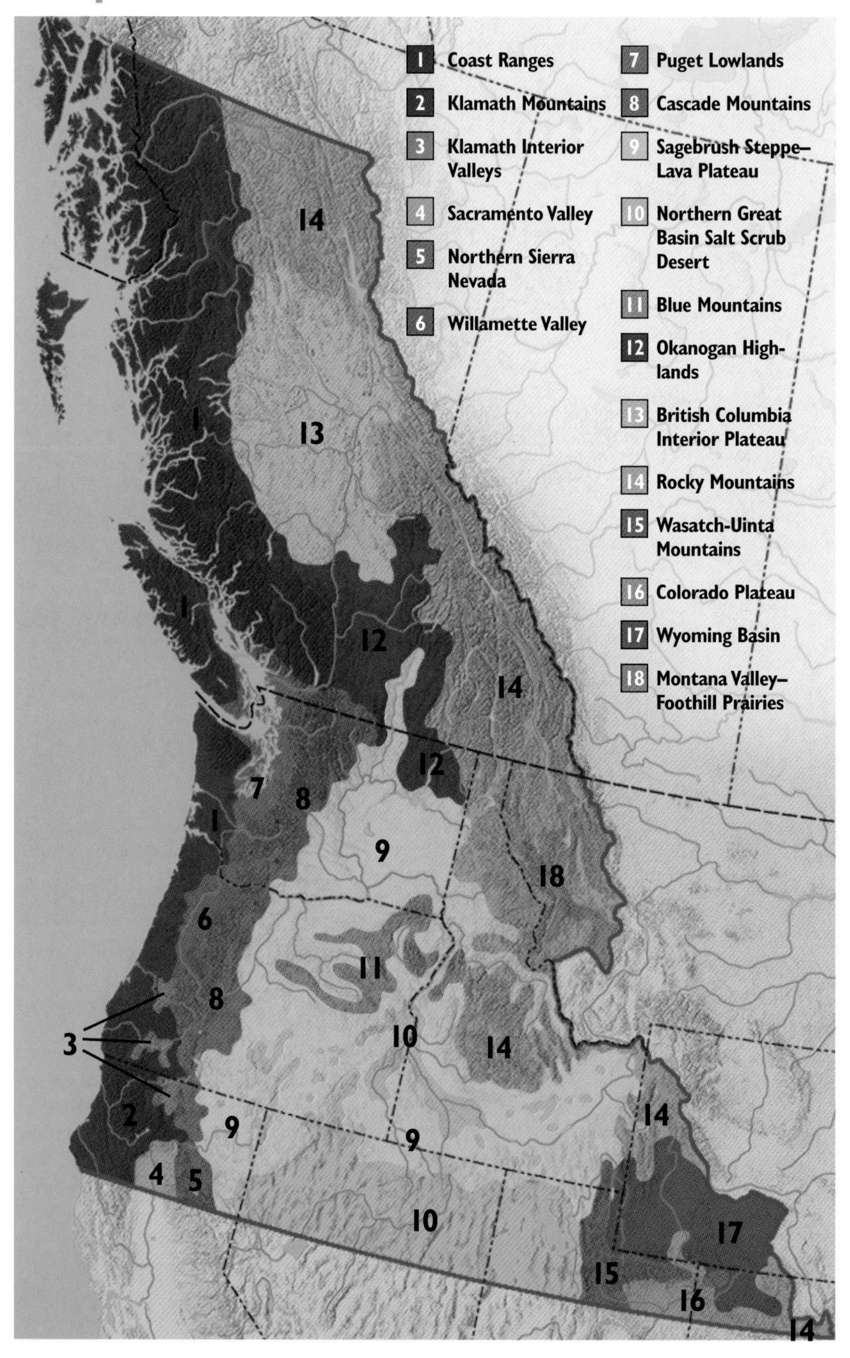

In several places, river drainages have breached these mountainous barriers, allowing some reptile species various degrees of dispersal east and west. These are the Pit and Klamath river canyons of northern California, the Columbia River Gorge along the Oregon/Washington border, and British Columbia's Fraser River Canyon. Northern expansion of reptile species into our region has been afforded by interlinking dry lake beds in the deserts of the Great Basin, in addition to the Colorado River drainage of eastern Utah and western Colorado. Similarly, low-elevation passes and interior valleys in the Klamath Mountains have permitted reptiles to move from California northward into the western portions of Oregon and Washington. Some reptiles typical of the Great Plains have spread west of the Continental Divide into the Northwest through lower passes in the Rocky Mountains along Montana's Clark Fork River drainage and the Great Divide Basin in Wyoming.

Of particular interest are several disjunct reptile populations found at considerable distances to the north of their species' primary distributions. Examples are the California Mountain Kingsnake in the Columbia River Gorge and the Sharp-tailed Snake on several islands in the Puget area of southwestern British Columbia. These relict populations are the result of gradual climate changes since the last glacial period. Approximately 4,000 to 6,600 years ago, there was a warmer, drier period, called an Altithermal or Hypsithermal Interval, that allowed many reptiles that require more arid, sunny conditions to range farther northward and higher in elevation than at present. When the region slowly reverted to a cooler climate, these reptiles retreated southward. However, some populations became stranded in isolated places that, for one reason or another, retained their requisite warm, dry habitat and allowed them to survive.

ECOPROVINCES

The following descriptions of the differing natural features found in the Northwest are intended to provide further clarification about the habitats mentioned in the reptile species accounts. I have broken the region into 18 ecoprovinces. They are based upon my own observations and the published work of a number of authorities in the fields of zoogeography and biogeography. I have attempted to create a synthesis of these various "physiographic regions," "bio-geo-climatic zones," "ecoregions," "life-zones" and other methodologies of placing a categorical framework upon the environmental complexity of the Northwest. My divisions may be an oversimplification in certain respects, but for the purposes of this book I think they will help the reader gain a better understanding of why each reptile species ranges where it does in the region. The map on the facing page illustrates the locations of these ecoprovinces. Because certain kinds of vegetation are often a useful indicator of a reptile's typical habitat, descriptions of the distinctive plant communities for each area are given in some detail. There is a listing on page 252 that provides the scientific names for the species of plants mentioned here and elsewhere in this book.

Coast Ranges. This ecoprovince extends from Oregon's Coquille River to northern British Columbia. The relatively low-elevation Coast Ranges of Oregon and southwestern Washington vary from sea level to 4,000 ft (1,220 m) and have a mild climate. To the

Grassy openings are good places to see Northwestern Garter Snakes and Northern Alligator Lizards in the mostly forested coastal areas. The Nature Conservancy's Cascade Head Preserve, northern Oregon coast.

north, the higher mountains of the Olympic Peninsula, Vancouver Island Range and British Columbia's Coast Mountains contain some peaks that rise to well over 8,000 ft (2,440 m) and have significantly cooler temperatures. Because of yearly rainfall averages that may exceed 150 in (380 cm) in places, the entire area is characterized by extensive coniferous rain forests that consist predominately of Douglas-fir, Western Hemlock, and Western Redcedar. A narrow band of Sitka Spruce occurs primarily on the often foggy seaward slopes. Although the shady, moist habitat of mosses and Western Sword Fern that prevails beneath these giant trees is conducive to a large array of amphibian species, few reptiles are found there. Only the Northern Alligator Lizard, Northwestern Garter Snake, and Common Garter Snake inhabit most of this ecoprovince.

Klamath Mountains. Located in the southwestern portion of our region, this ecoprovince is herpetologically fascinating. The rugged landscape was not heavily scoured by glaciers and contains some of the oldest and most complex geology in North America. This trait, coupled with mostly low elevations and a rather mild climate, has allowed an amazing diversity of flora and fauna to develop, including numerous endemic species. There is a narrow coastal strip in northern California and extreme southwestern Oregon that has a foggy climate with humid Redwood forests and a limited variety of reptiles. The interior, mountainous sections, however, have drier, sunnier conditions and more reptile species than are found in most other sections of the Northwest. The Klamaths are composed of several subranges, such as the Siskiyou Mountains, Marble Mountains, and Trinity Alps, and elevations range from sea level to somewhat over 9,000 ft (2,740 m). The Klamath Mountains are noted for having one of the most diverse coniferous forests in the world. Typical species are Ponderosa Pine, Sugar Pine, Knobcone Pine, Incense-cedar, Douglas-fir, White Fir, and Red Fir. On more open, south-facing slopes there is also an interlacing of broadleaf trees, notably Oregon White Oak, California Black Oak, Canyon Live Oak, Tanoak, Golden Chinquapin, and Pacific Madrone. An often thick chaparral undergrowth consists of such shrubs as White Manzanita, Buckbrush, Deerbrush, and Western Azalea. Scattered small bogs support insectivorous California Pitcher Plants. Reptiles that are more or less unique to this ecoprovince are the *shastensis* subspecies of the Northern Alligator Lizard,

Sagebrush Lizards are often common in the Klamath Mountain ecoprovince on exposed ridges with serpentine soils and scattered Jeffrey Pines. Kalmiopsis Wilderness, Siskiyou Mountains, southwestern Oregon.

The Illinois River Canyon in the Siskiyou Mountains of southwestern Oregon. Habitat of the California Mountain Kingsnake and Pacific Coast Aquatic Garter Snake.

the California Mountain Kingsnake, the *elegans* subspecies of the Western Terrestrial Garter Snake, the Pacific Coast Aquatic Garter Snake, and the newly described mountain forest form of the Sharp-tailed Snake. The *gracilis* subspecies of the Sagebrush Lizard is frequently common in more barren areas where serpentine soils allow only a thinly vegetated cover of Jeffrey Pine, White Manzanita, and various bunchgrasses.

Klamath Interior Valleys. Nestled within the Klamath Mountains, three valleys represent a middle ground between the relatively drier, more open Sacramento Valley to the south and the moist, lush Willamette Valley and Puget Lowlands to the north. These are Oregon's Umpqua Valley and Rogue River Valley (with its connecting tributary valleys of the Applegate and Illinois Rivers), and California's Shasta Valley (including the smaller adjoining Scott Valley and upper Klamath River drainage). Elevations range from

The Rogue River Valley viewed from The Nature Conservancy's Lower Table Rock Preserve, southwestern Oregon.

the Umpqua Valley's low of around 500 ft (150 m), through 1,000–1,900 ft (300–580 m) in the Rogue River Valley, to an average of about 2,400 ft (730 m) in the Shasta Valley. Coniferous forests of the surrounding mountains extend downward into these valleys in many places. The foothills and valley floors are predominantly more open, with grassy savannas that intermingle with woodlands of Oregon White Oak, California Black Oak,

Oak savanna in the Rogue River Valley, habitat of the Common Kingsnake and Sharp-tailed Snake.

and Pacific Madrone, along with a chaparral mix of Buckbrush, White Manzanita, and Western Poison Oak. On exposed, semi-arid southern slopes in the Rogue River Valley and Shasta Valley there is often an association of Western Juniper, Gray Rabbitbrush, and Big Sagebrush like that found to the east of the Cascade Mountains. These valleys have acted as corridors for southern species like the Common Kingsnake and Sharp-tailed Snake to extend their distributions northward, and for the Striped Whipsnake to range west of the Cascades.

An open hillside with rocks and mixed oak-juniper woodlands in the Shasta Valley of northern California. The Striped Whipsnake occurs west of the Cascade Mountains in this type of habitat.

Sacramento Valley. The northern end of this huge California valley just barely enters the region covered by this book. Significantly more open, dry, and warm than the Klamath Interior Valleys to the north, there are extensive grasslands on the valley's 400–700-ft (120–210-m) floor and lower foothills. The dominant trees are Blue Oak, Valley Oak, and Western Juniper, with California Sycamore along riparian zones. The higher surrounding foothills have Gray Pine mixed with many of the same species of trees and chaparral as found in the Klamath Mountains. This area is the only one in our region where the *mundus* subspecies of the Western Whiptail, the *multicarinata* subspecies of the Southern Alligator Lizard, the California Whipsnake, and the brown crossbanded variation of the Common Kingsnake are found.

Grasslands in the northern Sacramento Valley, California.

Blue Oak savanna on foothills at the eastern edge of the Sacramento Valley, near Red Bluff (The Nature Conservancy's Dye Creek Preserve). This warm, low-elevation ecoprovince has an exceptional diversity of reptiles, including the California Whipsnake.

Northern Sierra Nevada. Most elevational maximums are limited to under 8,000 ft (2,440 m) where these California mountains extend into the Northwest. An exception is Lassen Peak, at 10,457 ft (3,187 m). It should be noted that the Lassen area is situated at the northernmost tip of the Sierra and is sometimes considered part of the Cascade Mountains because the geology is similarly volcanic and the vegetation there is a blend from both ranges. South of Lassen, however, plant communities are more diverse and uniquely Sierran in composition, owing to differing geological and climatic conditions. Unlike in the Cascade and Klamath mountains just to the north, White Fir and Red Fir are considerably more dominant, and in many areas Ponderosa Pine is replaced by extensive stands of Jeffrey Pine. The western foothills are clothed in mixed oak-pine woodlands and chaparral. The eastern lower slopes of the Sierra consist of a semi-arid belt of more open pine forests,

The Susan River on the eastern slope of the Sierra, habitat of the Sierra Garter Snake.

along with Western Juniper, Big Sagebrush, Gray Rabbitbrush, Bitterbrush, and groves of Quaking Aspen. Occasional patches of Oregon White Oak and California Black Oak occur along the eastern base of the range at the edge of the Great Basin. The variety of reptile species found in this ecoprovince is much the same as in the Klamath Mountains. The beautiful California Mountain Kingsnake inhabits these forests and the Sierra Garter Snake and *pulchellus* subspecies of the Ring-necked Snake enter the Northwest here.

Willamette Valley. This broad river plain is a moister environment than the Klamath Interior Valleys to the south. The mild climate is due to the low elevations that vary from just above sea level at the north end, to about 600 ft (180 m) at the southern extremity. The nearby Pacific Ocean also has a strongly moderating influence on the weather. Originally, extensive grasslands (primarily created through millenniums of controlled burning by indigenous Indian tribes) carpeted the upland terraces, isolated buttes, and surrounding foothills of the valley. Within and around these wildflower-sprinkled prairies was a mosaic of scattered groves and woodlands of Oregon White Oak. Pacific Madrone and Western Poison Oak intermixed with the oaks, while Bigleaf Maple and Douglas-fir grew along riparian zones and on the cooler northern slopes. The bottomlands contained Oregon Ash swamps, immense marshes, and sphagnum bogs. Today, because of agricultural alterations, dams on the river, fire suppression, and urban growth, only fragments of this former landscape exist. Unfortunately, even these few less disturbed areas have been distorted by introduced exotic grasses, Scotch Broom, and other alien plants and animals. During the early part of the 20th century, Western Pond Turtles were still plentiful in aquatic habitats and Western Rattlesnakes occurred on most rocky southern exposures. At present, there are few breeding populations of these turtles, and rattlesnakes have been totally extirpated from all but the southern sections of the Willamette Valley. However, Western Fence Lizards, Southern Alligator Lizards, Western Skinks, Rubber Boas, Sharp-tailed Snakes, Ring-necked Snakes, Racers, Gopher Snakes, Northwestern Garter Snakes, Common Garter Snakes, and Western Terrestrial Garter Snakes are still found throughout many parts of the valley.

A rocky slope on a butte in the southern Willamette Valley, northwestern Oregon. Western Rattlesnakes occur in this habitat, along with Ring-necked Snakes, Sharp-tailed Snakes, and Western Skinks.

These open, grassy hillsides with scattered Oregon White Oaks, bordered by wetlands below, probably represent the original appearance of large portions of the Willamette Valley. Baskett Slough National Wildlife Refuge, near Dallas, northwestern Oregon.

A Puget Lowland prairie near Olympia, Washington. Northern Alligator Lizards, Northwestern Garter Snakes, Western Terrestrial Garter Snakes, and Rubber Boas occur within or around these openings.

Puget Lowlands. Situated in an elongated trough between the Coast Ranges and the Cascade Mountains, this ecoprovince extends north from the Columbia River through western Washington to extreme southwestern British Columbia. Included are the islands of Puget Sound, along with the southern and southeastern lowlands of Vancouver Island and its nearby smaller islands. Elevations range from sea level to about 600 ft (180 m), except for occasional higher promontories. The climate is mild and can be quite rainy, especially during the winter. There is a rainshadow created by the Olympic Mountains that allows somewhat drier conditions on many of the leeward islands. Habitats are more richly vegetated than in the Willamette Valley, with dense coniferous forests of Douglas-fir, Western Hemlock, and Western Redcedar in combination with Bigleaf Maple, Red Alder, Pacific Dogwood, and other deciduous trees. Grassy prairies with groves of Oregon White Oak exist in places with well-drained soils of glacially deposited sand and gravel. Most of these grasslands have disappeared under agricultural cultivation, urban development, and lack of fires. Oak and Pacific Madrone associations occur on south-facing rocky areas along the margins of Puget Sound, southern Vancouver Island, and the smaller offshore islands. Northern Alligator Lizards and Northwestern Garter Snakes are ubiquitous. Western Fence Lizards and Rubber Boas have patchy distributions (apparently absent on islands), and the Western Terrestrial Garter Snake primarily inhabits grassy oak habitats. The *pickeringi* subspecies of the Common Garter Snake is unique to the Puget Lowlands, and there are relict populations of the Sharp-tailed Snake on Vancouver Island and some of the adjacent smaller islands of British Columbia.

A seasonal grassy wetland near Tumwater, western Washington. The pickeringii *subspecies of the Common Garter Snake, unique to the Puget Lowlands ecoprovince, inhabits areas of this type.*

North Pender Island, with Saltspring Island beyond, typifies this archipelago of the Puget area. Relict populations of the Sharp-tailed Snake are found on some of these British Columbia islands.

Cascade Mountains. Volcanic in origin, the Cascades are the primary dividing mountains that create two entirely different west/east climates in the Northwest. These mountains extend north from the Pit River in northern California to the Fraser River

Canyon in extreme southern British Columbia. Prominent snow-capped peaks rise from the crest to high elevations. Notable are Washington's Mt. Rainier (14,411 ft/4,393 m), Oregon's Mt. Hood (11,235 ft/3,425 m), and California's Mt. Shasta (14,162 ft/4,317 m). The Cascade Mountains receive abundant rainfall on the western slopes, which results in dense coniferous forests of Douglas-fir, Western Hemlock, and Western Redcedar, intermixed with deciduous Bigleaf Maple, Vine Maple, Red Alder, and Pacific Rhododendron. The much drier and more open forests of the eastern slopes are predominately comprised at mid-elevations of Lodgepole Pine, along with Grand Fir in the northern sections of the range and White Fir, Red Fir, and Incense-cedar to the south. Extensive Ponderosa Pine forests are characteristic at lower elevations. In the higher zones of these mountains there are stands of Subalpine Fir, Silver Fir, Noble Fir, Engelmann Spruce, Mountain Hemlock, and Whitebark Pine. Because of shady forests and high elevations, reptile species in the Cascades are limited primarily to Northern Alligator Lizards, Rubber Boas, Common Garter Snakes, and Northwestern Garter Snakes. Western Fence Lizards, Western Skinks, and Ring-necked Snakes sometimes occur in lower canyons where there are rocky southern exposures. Similarly, the California Mountain Kingsnake ranges up canyons in the southern Cascades and the Columbia River Gorge. One of the few reptiles that inhabits the high Cascade crest (as far north as central Oregon) is the Pigmy Short-horned Lizard.

Sagebrush Steppe–Lava Plateau. Often referred to collectively as "High Desert," this immense ecoprovince is not a true desert. It takes in the Columbia Basin of eastern Washington, adjacent west-central Idaho, and north-central Oregon, the Snake River Valley and other sections of southern Idaho, and the northern fringes of the Great Basin in southeastern Oregon, northeastern California, and parts of northern Nevada and Utah. A narrow belt of sagebrush habitat that extends northward into south-central British Columbia is also included. Elevations range from around 98 ft (30 m) along the Columbia River at The Dalles, to over 9,000 ft (2,740 m) on some of the isolated mountain ranges that are found within the province. The diverse, semi-arid landscape is composed of a complex mosaic that includes plains of Big Sagebrush and Gray Rabbitbrush, bunchgrass prairies, lava fields, basaltic buttes, canyons, fault block escarpments, Western Juniper woodlands, and, where moisture allows, stands of Ponderosa Pine with a Bitterbrush understory. On some ranges that rise sufficiently enough to capture the required precipitation, there are forests of Douglas-fir, Subalpine Fir, Whitebark Pine, and Quaking Aspen. The Columbia Basin originally contained extensive grasslands

This sagebrush steppe landscape near Paisley in southeastern Oregon is typical of large portions of the interior high plateau country of the Northwest. Sagebrush Lizards, Pigmy Short-horned Lizards, and Gopher Snakes are commonly encountered in this habitat.

Basalt rimrock is seen throughout the Sagebrush Steppe–Lava Plateau ecoprovince and often harbors Western Fence Lizards, Racers, Striped Whipsnakes, and Western Rattlesnakes. Fort Rock Valley, central Oregon.

The eastern Columbia River Gorge along the Oregon/Washington border has oak woodland habitats. Reptiles that are more usually associated with valleys west of the Cascade Mountains occur there.

(Palouse Prairie), but most of this habitat has disappeared because of agricultural conversion. Typical reptiles native to this ecoprovince are Western Fence Lizards, Sagebrush Lizards, Common Side-blotched Lizards, Pigmy Short-horned Lizards, Western Skinks, Racers, Striped Whipsnakes, Gopher Snakes, Night Snakes, and Western Rattlesnakes. Common and Western Terrestrial garter snakes occur in aquatic areas, and Painted Turtles range throughout many of the Columbia Basin drainages. There are limited oak woodlands in northeastern California, south-central Oregon, north-central Oregon, the eastern Columbia River Gorge, and south-central Washington. Reptiles more associated with the interior valleys west of the Cascades, such as the Southern Alligator Lizard, Ring-necked Snake, Sharp-tailed Snake, and California Mountain Kingsnake occur in some of these oak habitats.

Northern Great Basin Salt Scrub Desert. This sun-baked land contains the most arid areas found within the Northwest. Sometimes called "basin and range country," this ecoprovince is dominated by ancient dry lake basins with dune systems and shimmering white alkali playas, interspersed by towering fault block ranges. Only the northernmost sections of the vast Great Basin extend into our region. It includes much of northern Nevada and northwestern Utah, with scattered pockets of this salt scrub habitat occurring in dry lake beds in northeastern California and southeastern Oregon. These arid shrublands are also found in the canyonlands of Oregon's Owyhee River drainage and in the adjoining Snake River Valley of Idaho, although these areas are technically not a part of the true, internally drained Great Basin. Elevations are relatively high overall, varying in Nevada from 3,880 ft (1,180 m) at Pyramid Lake, to over 11,000 ft (3,350 m) in the Ruby Mountains. An exception is along the Snake River on the Oregon/Idaho border, where the elevation drops to just over 2,000 ft (610 m). The salt scrub ecosystem has sandy, alkaline soils with

A salt scrub ecosystem in the Alvord Basin below Steens Mountain, southeastern Oregon. Sandy-gravelly areas adjacent to brushy dunes are excellent places to see Long-nosed Leopard Lizards, Desert Horned Lizards, and Western Whiptails.

Loose-soiled slopes with scattered boulders are the favored habitat of the Great Basin Collared Lizard. Pueblo Mountains, southeastern Oregon.

Several desert reptile species typical of the American Southwest reach the northern limits of their ranges in northwestern Nevada. The Pyramid Lake area is a good place to observe Zebra-tailed Lizards, Desert Spiny Lizards, Western Patch-nosed Snakes, and Coachwhips. Note: it is illegal to capture wildlife on this Paiute Indian Reservation.

desert shrub species including Big Greasewood, Shadscale, Spiny Hopsage, and Four-winged Saltbush, often intermixing with various sagebrush species at its fringes. A number of reptile species typical of the American Southwest deserts reach the northern limits of their ranges in this ecoprovince. These are the Great Basin Collared Lizard, Long-nosed Leopard Lizard, Desert Horned Lizard, Desert Spiny Lizard, Zebra-tailed Lizard, Western Whiptail, Coachwhip, Western Patch-nosed Snake, Long-nosed Snake, and Ground Snake. However, these heat-loving reptiles drop out where elevations on intervening mountain ranges graduate upward through sagebrush-juniper steppe, to pine-fir-aspen forests on the higher peaks.

Idaho's Snake River Valley has extensive areas of salt scrub desert, such as this example at Bruneau Dunes State Park. The northernmost limit of the Long-nosed Snake's range is in southwestern Idaho.

Blue Mountains. These 3,000–9,500-ft (910–2,900-m) mountains take in most of northeastern Oregon, extreme southeastern Washington, and a small portion of western Idaho. Several subranges comprise this ecoprovince: the Ochoco, Aldrich, Strawberry, Elkhorn, Greenhorn, Wallowa, Blue, and Seven Devils. In contrast to these high peaks, the 5,400-ft (1,650-m) deep Hells Canyon of the Snake River creates an enormous gorge between the Wallowa and Seven Devils

A number of isolated mountain ranges jut up from the Great Basin. Like desert-surrounded islands, their summits often harbor disjunct populations of forest-dwelling reptiles. Pictured are the Stansbury Mountains in northwestern Utah, where Regal Ring-necked Snakes and Milk Snakes occur.

ranges. Situated geographically between the Cascade and Rocky mountains, this province has characteristics of both ranges. Nevertheless, the Blue Mountains probably have more in common with the northern Rockies. There is a complex geological mix of very old sedimentary deposits and limestones, along with glacially eroded upthrusts of granite and more recent volcanic basalt. Plant communities of the interlacing valleys and lower slopes include sagebrush-rabbitbrush-juniper associations and bunchgrass prairies that grade upward into Mountain Mahogany with Ponderosa and Lodgepole pine stands. There is sufficient rainfall at higher elevations to support denser forests of Douglas-fir, Subalpine Fir, Western Larch, Whitebark Pine, Quaking Aspen, and Western Mountain Maple. Reptile variety is limited and largely confined to the lower, warmer

The various subranges of the Blue Mountains (here, the Wallowa Mountains in northeastern Oregon) are not especially rich in reptile life. Most species are confined to the foothills and adjacent valleys.

foothills. Expect to observe Western Fence Lizards, Sagebrush Lizards, Western Skinks, Rubber Boas, Racers, Gopher Snakes, Common and Western Terrestrial garter snakes, and Western Rattlesnakes.

Okanogan Highlands. This ecoprovince encompasses most of northeastern Washington and south-central British Columbia and is characterized by river valleys and rather flat-topped hills and mountains. Composed of sedimentary deposits and rocks of volcanic origin, most elevations are at 3,000–5,000 ft (910–1,520 m), with some higher prominences of slightly over 7,000 ft

Okanagan Lake in the Okanagan Valley of south-central British Columbia. This dry, interior area marks the northern limits for several sagebrush steppe reptiles: the Racer, Gopher Snake, Night Snake, Western Rattlesnake, and possibly, the Pigmy Short-horned Lizard.

(2,130 m). Stands of Ponderosa Pine with Big Sagebrush, Gray Rabbitbrush, and bunchgrasses are found in valleys and on lower slopes. At mid-elevations there are thick forests of Lodgepole Pine, Douglas-fir, White Spruce, and Quaking Aspen, with Subalpine Fir and Engelmann Spruce on higher mountains. A narrow, semi-arid belt of lower-elevation, open sagebrush habitat extends northward into British Columbia along the Okanogan River drainage. That area has been included as part of the Sagebrush Steppe–Lava Plateau ecoprovince. Painted Turtles, Western Skinks, Northern Alligator Lizards, Rubber Boas, and Common and Western Terrestrial garter snakes have been recorded throughout many parts of the Okanogan Highlands. Racers, Gopher Snakes, and Western Rattlesnakes are primarily confined to the southern and western portions of the ecoprovince.

British Columbia Interior Plateau. Primarily located in the upper Fraser River drainage, this northern landscape takes in most of interior central British Columbia between the Coast and Rocky mountains, north of the Okanogan Highlands. The rolling elevations of plateaus, plains, and low mountains generally do not exceed 3,000–5,500 ft (910–1,680 m). The dominant habitat is one of forests, consisting of White Spruce, Douglas-fir, Lodgepole Pine, Paper Birch, and Quaking Aspen on lower slopes, with Subalpine Fir, Engelmann Spruce, and tundra at the higher zones. There are grasslands along the river bottoms, and Black Spruce bogs are common in the northern parts of the area. Summers are short and the winters prolonged, resulting in a scarcity of reptiles. Only Common and Western Terrestrial garter snakes extend their distributions north into this ecoprovince to any extent.

Rocky Mountains. These formidable mountains form the eastern boundary of our region where the Continental Divide snakes along the rugged crest from British Columbia to northern Colorado. Many peaks exceed 13,000 ft (3,960 m), which results in harsh climates with short warm seasons, particularly in the Canadian Rockies. Relatively more temperate conditions are to be found farther south along the range in Colorado. Coniferous forests cover these mountains but vary in species make-up depending upon the location. The western slopes of the northern Rocky Mountains receive more rainfall and have dense, mid-elevation stands of Douglas-fir, Western Redcedar, Western Hemlock, Western Larch, and White Spruce, along with Western Mountain Maple and Quaking Aspen. On the higher areas are Subalpine Fir, Engelmann Spruce, and Whitebark Pine. At lower elevations, on the eastern slopes, and in the southern sections of the Rockies, conditions are drier, with more open forests of Lodgepole Pine, Ponderosa Pine, and Rocky Mountain Juniper. In far northern British Columbia these mountains gradually recede to moderate elevations with boreal forests of mixed conifers and deciduous trees. Rubber Boas and Common and Western Terrestrial garter snakes are among the few reptiles that range upward into the Rocky Mountain forests, sometimes to more than 7,000 ft (2,130 m) in elevation south of Canada. Other reptile species from surrounding ecoprovinces, such

A meadow at the border of the Mt. Timpanogos Wilderness in the Wasatch Range of northwestern Utah, typical habitat for the Smooth Green Snake. Milk Snakes, Sonoran Mountain Kingsnakes, and the Regal Ring-necked Snake also occur in these mountains.

as Sagebrush Lizards, Racers, Gopher Snakes, and Western Rattlesnakes occur in the sagebrush and grassland habitats of the foothills and adjoining valley margins.

Wasatch-Uinta Mountains. Although considered to be isolated components of the Rocky Mountains, these two ranges are unique in several ways and are treated here as a separate ecoprovince. The Wasatch Range extends north through the entire length of central Utah into extreme southeastern Idaho. The Uinta Mountains are located in northeastern Utah and a small portion of adjacent northwestern Colorado. It is one of the rare mountain ranges in North America that run in an east-west direction. Situated at the eastern edge of the arid Great Basin, the climate is significantly drier than other parts of the Rocky Mountains owing to the far-reaching rainshadow of the Sierra Nevada in California. Some peaks exceed 11,000 ft (3,350 m) in the northern Wasatch Range and rise to over 13,000 ft (3,960 m) in the Uintas. Mid-elevation forests consist of Ponderosa Pine, Limber Pine, and Douglas-fir, with Subalpine Fir, White Fir, Engelmann Spruce, Western Mountain Maple, and Quaking Aspen at the higher zones. The lower slopes of these mountains are more open, with grassy areas and extensive patches of scrubby Gambel Oak, along with Canyon Maple, Utah Juniper, and Two-needle Pinyon Pine. The Wasatch Range and its adjoining valleys and lesser mountains is the only place within our region that the *regalis* subspecies of the Ring-necked Snake is found, and the Sonoran Mountain Kingsnake has the northernmost limits of its distribution there. Additionally, the Milk Snake and Smooth Green Snake reach the western extent of their ranges in both the Wasatch and Uinta mountains.

Colorado Plateau. Only the northern tip of this classically American Southwest landscape enters the area covered by this book in the Green, Duchesne, and Yampa river drainages of northeastern Utah, northwestern Colorado, and an extremely small section of southwestern Wyoming. Typical Colorado River sandstone formations are found there, although not as large and dramatic as those

Sandstone formations and a pinyon-juniper "pigmy forest" at Red Fleet State Park in the Uinta Basin of northeastern Utah, habitat of the Eastern Fence Lizard, Ornate Tree Lizard, and Midget Faded Rattlesnake.

farther south in the famed national parks. Also to be seen is the semi-arid "pigmy forest" of Two-needle Pinyon Pine and Utah Juniper mixed with open areas of Big Sagebrush, Gray Rabbitbrush, and bunchgrasses, along with a variety of cacti and yucca species. In places, there are arid areas of Great Basin salt scrub species, including Big Greasewood, Shadscale, Four-winged Saltbush, and Spiny Hopsage. Although the overall elevations are high with most valley bottoms in excess of 5,000 ft (1,520 m), there is a good variety of reptile species native to this ecoprovince. Of special interest are the Eastern Fence Lizard, Ornate Tree Lizard, Greater Short-horned Lizard, Smooth Green

Snake, Corn Snake, Milk Snake, and Midget Faded Rattlesnake (a subspecies of the Western Rattlesnake).

Wyoming Basin. Located in southwestern Wyoming and a portion of northwestern Colorado, this semi-arid to arid, open terrain forms a break in the usually lofty barrier of the Rocky Mountains. There, the Continental Divide meanders across the 6,000–7,000-ft (1,830–2,130-m) Great Divide Basin, which has no drainage to any ocean. This is one of the few east-west routes for a number of terrestrial animal species to move between the Great Plains and the Intermountain West. The topography varies from sagebrush steppe, grasslands, and sand hills, to eroded badlands in Wyoming's Red Desert. Arid sites often have a growth of salt scrub, such as Big Greasewood and Shadscale. On higher plateaus and isolated mountains there are scattered Utah Juniper with Bitterbrush, graduating upward to patches of Gambel Oak, Quaking Aspen, and pine woods. Reptiles that have been recorded in this ecoprovince are the Eastern Fence Lizard, Sagebrush Lizard, Ornate Tree Lizard, Greater Short-horned Lizard, Racer, Milk Snake, Smooth Green Snake, Gopher Snake, Western Terrestrial Garter Snake, and prairie subspecies of the Western Rattlesnake.

Two Great Plains reptile subspecies, the Bullsnake and Prairie Rattlesnake, range into western Montana. Both can often be seen in prairie areas, such as the foothill habitat pictured here at the National Bison Range.

Montana Valley-Foothill Prairies. These open, grassy habitats extend from the eastern side of the Continental Divide in the area of the Rocky Mountain Front and the upper Missouri River, over low passes through the Clark Fork River drainage and into the valleys of western Montana. They have provided an avenue of dispersal for two Great Plains reptiles to range into our region in Montana's Bitterroot and Flathead valleys: the Bullsnake (a subspecies of the Gopher Snake) and the Prairie Rattlesnake (a subspecies of the Western Rattlesnake). Although much of this ecosystem has been lost to agriculture, fire suppression, and other human alterations, there are a number of relatively natural areas remaining on the 2,900–3,500-ft (880–1,070-m) valley floors and foothills. Bluebunch Wheatgrass, Idaho Fescue, and other grass species grow there, along with areas of Big Sagebrush. In the valleys there are Black Cottonwood in riparian areas with scattered Quaking Aspen groves and stands of Ponderosa Pine. Denser forests of Lodgepole Pine and Douglas-fir cover the higher slopes. Painted Turtles, Western Skinks, Northern Alligator Lizards, Rubber Boas, Racers, and Common and Western Terrestrial garter snakes also occur within or at the fringes of this ecoprovince.

Grassy sand hills in the Cedar Springs Wildlife Area of northwestern Colorado. The Greater Short-horned Lizard, Milk Snake, and Prairie Rattlesnake occur in this section of the Wyoming Basin ecoprovince.

Observing Reptiles in the Field

Although there is still need for improvement, in recent years there has been a noticeable change for the better in our attitude toward reptiles. Possibly, this change is because of the popularity of exciting television documentaries about reptiles, along with magazine and newspaper stories that present these animals in a positive way. More enlightened school science programs are probably another influential factor. Additionally, exotic, captive-bred reptiles have become extremely popular as cage pets. Reptiles are in!

Nevertheless, there is a downside to this fascination with giant tropical constricting snakes and South American Green Iguanas. Our native Northwest reptiles are often overlooked by the general public. There is usually little knowledge about the various Northwestern reptile species in their own backyards and surrounding countryside. Environmental protectionists might argue that this lack of interest in our native reptiles is a good thing. "Don't encourage the hoards to go out looking for turtles, lizards, and snakes. Leave well enough alone!" is their advice. This reasoning certainly has its appeal. Unfortunately, urban sprawl, logging in pristine areas, and agricultural conversion eat up more and more natural habitat yearly. Inappropriate use of herbicides and pesticides, destruction of wetlands, and introduced non-native species of plants and animals continue to disrupt ecosystems. Small reptiles, which are sometimes overlooked in conservation efforts, need all the advocates they can get.

Bird-watching is one of the most popular outdoor pursuits, resulting in an intensely dedicated lobby on behalf of these feathered creatures. Consequently, there are far more refuges, population monitoring programs, and research projects for birds than reptiles. More reptile-watchers in the region would probably bring about similar results. However, there are some problematic differences that arise. Although many diurnal reptiles are often out moving about and can, like birds, be observed with binoculars, others are secretive. Searching under objects is required, and good field techniques must be used or we will destroy what we seek to protect and enjoy.

FIELD STUDY TECHNIQUES

Rule number one is "Do no harm." If there is any possibility that your reptile watching may damage the habitat, directly hurt the animal, disrupt breeding activity, interfere with feeding, or place undue stress on the population as a whole, back off. It's not worth the risk.

When scrupulous care is taken in turning over rocks, logs, bark, leaf-litter and other surface objects that might conceal reptiles, it is usually difficult to detect afterwards that any significant disturbance has taken place. In areas inhabited by venomous snakes, always tilt the object up toward yourself so that it is between you and whatever might be beneath, thus avoiding a possible bite. When a reptile is found, it is best to remove the animal first if

Gary Winter finding a juvenile California Mountain Kingsnake under a rock. He tilted the stone toward himself, using it as a shield in case there had been a rattlesnake hidden beneath. Columbia River Gorge National Scenic Area, Washington.

A Gopher Snake and two Ring-necked Snakes were found under this piece of sun-warmed tin when it was turned by Richard Hoyer. Wilson Wildlife Area, near Corvallis, Willamette Valley, northwestern Oregon.

replacing the object might crush it. After you have finished observing the reptile, allow it to escape back under the reseated object through any available crevice that leads into the retreat. A seal of sorts usually has developed around most of the edges of a rock or log that is sunk into the soil and has not been disturbed for a length of time. I always give the object a firm kick with my boot after replacing it, helping reset things back into the original resting place. If any smaller overlapping rocks, sticks, pieces of bark, or dead leaves have been pushed away from the object's borders, I try to replace them as closely as possible to their former configuration. When a rotting log looks so fragile that it might disintegrate when being shifted, or if a large rock is precariously balanced on a steep slope with the appearance that it might slide or roll down the hillside, pass it by. Old homesteads with scattered boards and pieces of roofing paper, tin, and plywood are also excellent places to find hiding reptiles, but carefully replace these objects, too.

For those people who lack the inclination, time, or ability to perform the more industrious field work previously described, there are other less strenuous aspects to reptile-watching. As mentioned before, a number of active, diurnal species can often be observed with binoculars or even approached close enough to see detailed behavior with the unaided eye. The most profitable times for these outings are generally in the spring and autumn, or during summer mornings and evenings when temperatures are more moderate. Search sunny, south-facing slopes, particularly where there are rock ledges, jumbles of stones, logs, or stumps that provide basking sites. Watch for lizards and snakes carefully, because they are often partially concealed or have protective camouflage that blends in with the habitat remarkably well. Continued practice sharpens the eyesight for this sort of

Morning in the desert is a good time to see basking lizards and snakes on rocky slopes. Big Basin, southeastern Oregon.

wildlife-watching. In deserts, walking slowly between shrubs on sandy flats, dune systems, and hillsides will frequently turn up the large Great Basin lizard species, such as Long-nosed Leopard Lizards, Desert Horned Lizards, and Western Whiptails. Great Basin Collared Lizards are encountered on loose-soiled slopes with scattered boulders. Turtles are extremely wary and have keen eyesight, so binoculars are a must. Look for them basking on open

The large and active Desert Spiny Lizard is a species easily observed through binoculars. Black Rock Desert, northwestern Nevada.

Walking slowly between desert shrubs in sandy areas will often turn up Long-nosed Leopard Lizards, Desert Horned Lizards, and Western Whiptails. Big Sand Gap, Alvord Desert, southeastern Oregon.

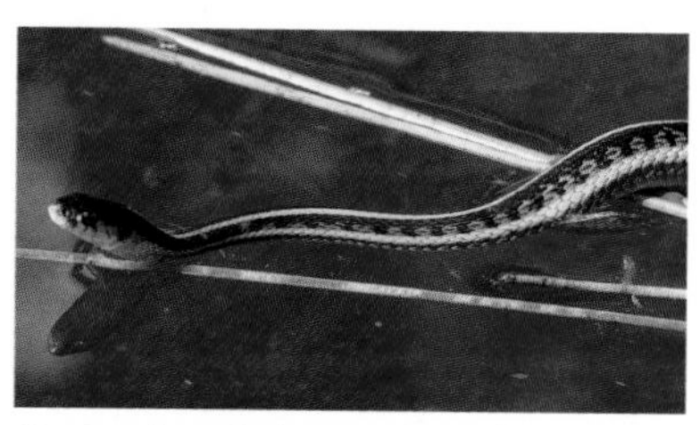

Common Garter Snakes are often seen around aquatic areas, frequently earning them the erroneous name of "water snake." Crane Prairie Reservoir, Cascade Mountains, Oregon.

banks, floating logs, and cattail mats along slow stretches of streams and rivers or in ponds and lakes. Use screening shoreline vegetation as cover for a closer approach. If the turtles detect your presence and dive into the water, remain motionless for a few minutes and they will usually return to their basking sites. Several species of garter snakes also occur in these same aquatic habitats, where you may have the chance to observe one eating a frog or tadpole.

For other suggestions about reptile-rich environments, see the information under "Habitat and Behavior" in the species accounts and the section beginning on page 27, titled "Reptile Habitats in the Northwest." The "Field Notes" sections that accompany each species account provide additional reptile-watching tips. In these stories I have frequently included the names of good locations where various reptile species can be observed. The captions to the reptile and habitat photographs also give similar information. I have highlighted protected wildlife refuges, parks, nature preserves and other places that are open to the public. These areas often have good trail systems, interpretive information, and sometimes resident biologists and naturalists that can offer assistance. Appropriate trails for the handicapped are provided at some of these locations.

Ideally, when observing reptiles in the field it is best to just look and not touch. However, it is sometimes necessary to capture and handle a specimen to be sure of identification, or to make a photographic record of ventral markings and other details of its anatomy. Many lizards are so speedy that they are extremely difficult to capture by hand. Noosing is the most effective means of catching these elusive reptiles. Make a small slipnoose from fishing line and tie it to the end of a 4–6-ft (1.2–1.8-m) pole. A telescoping fishing rod is ideal, but if need be, a makeshift long stick scavenged from the surrounding environment will suffice. Slowly approach a lizard as closely as possible without scaring it away, stretch out the pole and carefully work the noose over the lizard's head until it is around the neck. A quick jerk upward with the pole and the specimen is noosed. This does not harm the lizard, and from my observations, causes little, if any, trauma. I once noosed a Common Side-blotched Lizard, restrained it in one hand while I photographed the belly markings, and then released it back to the rock

Naturalist Laura McMasters demonstrating the proper method of picking up a snake by the tail. She has grasped this Common Garter Snake well up toward the body to give more support to the reptile. Near McMinnville, Willamette Valley, northwestern Oregon.

Warm summer evenings and nights in the desert are a productive time to "road hunt" for crossing snakes, particularly just before or after a thunderstorm. Winnemucca Lake Basin, northwestern Nevada.

it was originally basking on. Immediately, the small reptile caught and ate a fly. Obviously, the lizard was not bothered in the least by its brief capture.

During the hot summer months when midday temperatures soar (particularly in deserts), many species of snakes become active only in the cooler mornings and evenings, or after dark if the night is warm enough for reptilian activity. The paved surface of a road will retain heat for several hours after sunset, even when the air has become chilly. Nocturnal snakes commonly lie on road surfaces and soak up the warmth. Herpetologists take advantage of this habit by practicing a search method called "road hunting," which entails slowly driving along roads that pass through good snake habitat. It can be a very productive way of seeing snakes that normally are rarely encountered. The "Field Notes" section that accompanies the Common Kingsnake species account (p. 193) gives more details about this particular field technique. Rattlesnakes are frequently found crossing roads, so when you are in their territory, it is a good idea to have what is usually called a "snake stick" or "snake hook," which is a 36–48-in (90–120-cm) staff (wood or metal) that has an angled metal hook attached to one end. It looks similar to a shepherd's hook and is used to lift venomous snakes and move them about. I often use mine to hook a rattlesnake to the roadside to prevent it from being run over by a passing auto (children, of course, should never attempt to capture a venomous snake.) Harmless snakes can be carried off the road to safety by hand. Snake sticks are also handy for poking in crevices or bushes and turning

Craig Zuger noosing a Great Basin Collared Lizard. Northwestern Nevada.

Fort Rock State Park in central Oregon is a good place to see sagebrush steppe lizard species, such as Pigmy Short-horned Lizards, Sagebrush Lizards, and Common Side-blotched Lizards.

Craig Zuger using a snake hook to move a Western Rattlesnake. Southern Willamette Valley, northwestern Oregon.

leaf and bark litter when hunting for reptiles, along with making an excellent walking staff when hiking. In regard to road hunting, it should be noted that dead reptiles are frequently encountered. If you find one that might constitute a range extension or is unusual in some other way, jot down some locality data and save the road casualty in an ice chest to be later given to a university or museum. These are valuable as voucher specimens.

Cloth "snake bags" are the best way to transport reptiles. Jan St. John with a bagged Western Rattlesnake, near Sisters, central Oregon.

If a captured reptile is to be kept for a longer period of time before releasing it, cloth "snake bags" are much better than bottles, jars and other such rigid containers that are difficult to carry into the field. These bags can be made from inexpensive muslin by anyone familiar with the use of a sewing machine. Old pillow cases will work, too, but inspect them carefully for holes and weak seams. Once a reptile is bagged, tie the top into a tight knot or twist it closed and secure it with several winds of a sturdy rubber band. Before leaving on an outing, simply tuck a snake bag through your belt or stuff it in a daypack and you are prepared.

PROTECTED SPECIES AND OTHER CONSERVATION ISSUES

Nature study is not as simple and straightforward as it once was. Some populations of reptile species are protected by state, provincial, and federal laws for a myriad of reasons, including habitat degradation, over-collecting, or concern for isolated occurrences of reptiles in fragmented, fragile habitats. In light of this, I must strongly emphasize that all my previous recommendations about field methodologies that involve capturing a reptile for close examination in the hand do **not** apply to the species protected by these laws. Please leave them alone.

Before going reptile-watching, first contact your local wildlife division office about protected species and the restrictions that are involved. Do the same with other wildlife divisions when you travel to different states

The eastern section of the Columbia River Gorge National Scenic Area, with its rocky, low-elevation oak woodlands, has an abundance of reptiles. The Washington canyon pictured has a population of the beautiful California Mountain Kingsnake. This species is protected in Washington and should not be captured or removed from its habitat.

and provinces to look for reptiles. I had considered giving a listing in this field guide of the status for each protected reptile species in the Northwest, but these regulations and their applications vary greatly with each state and province. Additionally, there are periodic changes in these restrictions that would soon render obsolete any listing provided. Also keep in mind that state and national parks and monuments, wildlife refuges, nature preserves, Indian reservations and other such designated areas usually do not allow the handling and collecting of any animals. Therefore, I will leave it to the reader to take personal responsibility of learning about and abiding by current laws. Capturing and taking protected species requires applying for a scientific collecting permit, which is only granted for legitimate purposes, such as university research projects, exhibit specimens for museums and zoos, educational purposes (including detaining protected reptiles for the photography needed in books like this one) and similar pursuits. In California, a fishing license is required to capture even the most common reptiles that are not on the protected species list. A copy of their state fishing regulations will provide details on the number allowed to be taken yearly per species. The state of Washington requires a scientific collection permit to take **any** species of reptile.

There is also the issue that protected species laws do not cover the protection of these reptile's habitats. Therefore, it is a matter of individual conscience and integrity that we police ourselves and use the highest standards of non-damaging field methods. Otherwise, a few "bad apples" will spoil things for everyone and laws will become increasingly stringent. Habitat protection is ultimately the most important form of conservation, because not just one species is preserved, but the entire complex of interdependent plants, animals, and other components that occur there. If you see habitat destruction or witness the collection of protected species, report it to local authorities immediately.

U.S. Forest Service biologist Robin Dobson examining habitat destruction caused by snake hunters in the Columbia River Gorge National Scenic Area. A large slab of rock had been pried from the ledge and shoved down the slope. Numerous rocks and logs had been similarly disturbed in the area.

Finally, another factor that relates to conservation is the keeping of reptiles as pets. This activity can be a positive educational experience for children that imparts lessons on responsibility, especially if overseen by a teacher in a classroom situation. Caring for these animals full-time is often more complicated than it may first appear. Such foods as slugs, insects, or mice may be difficult to consistently acquire, especially in the winter. For example, horned lizards require certain kinds of ants to eat and never live long in captivity. It is probably best to keep just one to three specimens of common, non-protected local species for a few days or weeks in the spring and summer and then release them before the arrival of cooler autumn weather. Any animal that refuses to eat should be freed before it becomes weakened. Always return the creature to the exact location of initial capture so that the proper habitat is assured.

As the popularity of enjoying our native reptiles increases, though, populations of even the most common species may begin to decline if too many are taken from the wild. It is better that captive-bred turtles, lizards, and snakes be purchased from reputable dealers. These reptiles are accustomed to living in cages and usually thrive when maintained properly. There are many excellent publications available pertaining to the care and breeding of reptiles.

Spectacularly colored, captive-bred snakes and other reptiles are available to purchase in pet shops or directly from breeders. This Red Albino Corn Snake is one example among many varieties. They thrive in captivity and are a better choice for pets than captured wild native species.

Field Notes. 27 April 1997. This day's field trip turned out to be very disheartening when I observed an instance of habitat abuse in the Columbia River Gorge National Scenic Area. A Washington side canyon there is well known among snake collectors for having a population of the highly desirable California Mountain Kingsnake (protected in that state). Due to its notoriety and relative proximity to Portland and Seattle, this particular location receives a concentration of attention from herpers that far exceeds that given to the more obscure sections of this species' range in the Northwest. This is particularly the case during early spring, because reptile hunters know that snakes are still grouped around hibernacula areas.

The weather was pleasantly warm as I hiked along my usual route that passes several rocky, south-facing denning sites used not only by California Mountain Kingsnakes, but by Western Rattlesnakes, Ring-necked Snakes, and other kinds of snakes, as well. I soon noticed that someone had preceded me, probably the day before. Everywhere I looked was devastation. Rocks were left overturned or rolled down slopes, along with numerous rotting logs and stumps that had been completely pulverized. Worst of all, large, flat stones situated directly within hibernacula were totally jumbled and left tilted up on edge in many instances. Snakes use these slabs to coil beneath on the first, barely warm spring days. The thin rocks quickly soak up solar rays and efficiently transmit heat to the reptiles. This solar warming aids snakes in gaining sufficiently high body temperatures to capture, swallow, and digest their first meals of the year and, in some instances, to mate. Even if the collectors failed in their attempt to capture Mountain Kingsnakes, more damage was probably done to this population by habitat disruption than if they had removed a dozen or more specimens. I did my best to repair the destruction and hiked back to the trailhead in a disgusted state of mind. I promptly reported these abuses to the U.S. Forest Service, which manages the area.

Caution: Never release non-native pets into local habitats. If they become established, exotics can displace and sometimes prey upon local species, or possibly introduce new diseases. In some states and provinces there are laws that prohibit the release of non-native animals.

THINGS YET TO BE LEARNED

One of the nice things about the wondrously complex web of life on our planet is that humanity will probably never learn everything about it all. Those of us who are fascinated by the natural world should never have reason for boredom. There is still an enormous amount to be discovered, and not just by professional scientists. The amateur "weekend naturalist" can contribute greatly to our store of knowledge by making accurate observations during his or her spare time. This is certainly the situation in the field of herpetology. Many species are extremely secretive and little is known about their life histories. Indeed, in the Northwest such basic information as the geographic ranges of many reptiles is imperfectly documented. A quick glance through the distribution maps in this book will prove this by the frequent question marks indicated here and there. There are a number of references made in the species accounts about questionable past locality records that are in need of verification, as well as many taxonomic, behavioral, and reproductive quandaries. The examples are numerous and will keep enterprising herpetologists busy for many years.

An evening storm over the vast Black Rock Desert playa in northwestern Nevada. This remote, harshly arid area of the northern Great Basin is poorly known herpetologically and needs further field study.

Biologist Christian Engelstoft weighing and recording data on a juvenile Sharp-tailed Snake. The information was added to an ongoing study of this species on British Columbia's Gulf Islands and southern Vancouver Island.

Distributional Challenges

- Reports of the California Mountain Kingsnake on the Oregon side of the Columbia River Gorge and in Washington's Yakima River drainage need to be verified.
- Does the Long-nosed Snake inhabit southeastern Oregon?
- In Utah's Wasatch Mountains, does the Smooth Green Snake range north of Ogden, and possibly into southeastern Idaho?
- Likewise, what is the northernmost limit for the Sonoran Mountain Kingsnake in the Wasatch Mountains, and is it also found in some of the mountain ranges of northeastern Nevada?
- Does the Zebra-tailed Lizard occur between its main distribution in western Nevada and the seemingly isolated population to the north in the Black Rock Desert?
- Are there relict populations of the Long-nosed Leopard Lizard in the Columbia Basin?

A Western Pond Turtle with a freshly attached transmitter. The reptile's movements were tracked with telemetry equipment for a survey of the species in Oregon's Willamette Valley.

- Do the Rubber Boa and Corn Snake inhabit northwestern Colorado?
- The 1964 record for the Sharp-tailed Snake in south-central British Columbia needs updated verification.
- Are there Desert Horned Lizards and Ground Snakes in northeastern Utah?
- Do Ground Snakes occur between their primary range in Nevada and the populations in the Snake-Owyhee drainages of Idaho and Oregon?
- Do the distributions of the Coast Horned Lizard (*Phrynosoma coronatum*) and the Giant Garter Snake (*Thamnophis gigas*) extend northward in the Sacramento Valley into our area?

There is an additional distributional puzzle that is of considerable interest. Three Western Hog-nosed Snakes (*Heterodon nasicus*) were collected during 1989 and 1990 in sand hills near the small community of Maybell, in Moffat County of northwestern Colorado. These records are the only ones for this species west of the Continental Divide and north of New Mexico. Despite several recent searches by the author and other naturalists and biologists, no more specimens have been found there. The big question is, were the snakes introduced, or do they represent an isolated natural population and an additional species for the Northwest?

Taxonomic Questions

- What are the relationships of the various color/pattern variations of the Northern Alligator Lizard in the Klamath Mountains and the isolated mountain ranges of the northern Great Basin?
- Is the recently discovered mountain forest form of the Sharp-tailed Snake a new species?
- How closely related are the morphologically similar Striped Whipsnake and California Whipsnake?
- Should the Corn Snake populations in the Colorado Plateau region be considered a separate subspecies?
- What is the significance of the divergently patterned Milk Snakes in Utah's Stansbury Mountains and the Western Terrestrial Garter Snakes in Oregon's Willamette Valley?
- Is it correct to assess the area where the ranges of the Pacific Coast Aquatic Garter Snake and Sierra Garter Snake overlap in California's Pit River drainage to be a zone of hybridization?

Behavioral and Reproductive Quandaries

It is safe to say that nearly all the species of reptiles native to the Northwest need every aspect of their life histories studied in greater detail to one degree or another. The species accounts in this book repeatedly state that little is known about a reptile's feeding habits, annual activity, hibernation periods, timing of breeding seasons, courtship, egg-laying/birthing behavior, and defenses against enemies. More marking programs (each reptile is tagged in some way for recognition) need to be undertaken over long periods that follow both the individual and overall population dynamics. Meaningful data on trends in population numbers, percentages of both sexes, sizes of individual home ranges, activity patterns, measured growth of the reptiles, mating habits, and other factors can be collected through these in-depth studies. Of particular importance in this era of environmental degradation and loss is to acquire more knowledge about the needs of each reptile species so that informed choices can be made for setting aside sanctuaries of healthy natural areas. In this way, representative habitats that protect the full range of reptiles (and other associated flora and fauna) native to each ecoprovince in the Northwest are preserved.

The most important first step is to record your observations, which can be as simple as having the fun of jotting down a quick notation in your "lifelist" when you see a reptile species that is new to you. In later years you will enjoy reliving memories of the first times you observed various kinds of turtles, lizards, and snakes in the wild by reading back through your list. A lifelist section is provided at the rear of this book where each species, subspecies, and significant geographical variation of the native Northwest reptiles is included. Headings of "Date," "Location," and "Remarks," with spaces to write brief entries are provided. I encourage going beyond life listing, though, and to also begin keeping a more lengthy journal to accurately record detailed observations concerning reptiles (and other aspects of nature) whenever you venture into the field. Doing occasional sketches in pencil or pen-and-ink will clarify your observations, and some naturalists include a compact kit of watercolor paints or colored pencils in their daypack to further enliven the pages of their field journals. The information you gather can be shared with the biology or zoology departments of a local university, a natural history museum, the nongame programs of wildlife divisions, wildlife heritage programs, The Nature Conservancy, and similar institutions and organizations with herpetological components to their work. They also may need volunteers for research projects, refuge work crews, educational outreach, or cataloging programs that involve reptiles. Who knows? Your field surveys and resultant journal notes might provide the locality data for a range extension, the description of previously undocumented behavior, or even the discovery of a new kind of reptile!

PHOTOGRAPHING REPTILES IN THE FIELD

Besides being a rewarding hobby, photographing turtles, lizards, and snakes is invaluable for visual documentation to compliment journal notes. It is also of benefit if a reptile is found that cannot be identified. Photos of the specimen can later be shown to a knowledgeable herpetologist. Additionally, the discipline of photographing the natural world (and sketching and painting) hones your powers of observation concerning details, subtle colors, textures, and beauty of form.

This book is not the place for involved instructions on the rudiments of photography, nor the overall subject of nature photography. There are a number of excellent books that thoroughly cover these basics (several are listed in the "Resources" section of this field guide.) Magazines like *Outdoor Photographer* are also good places to learn more. I will assume that the reader is at least somewhat familiar with photographic terminology, equipment, and techniques. My remarks here will be specifically limited to a number of practical tips for making quality reptile photographs in the field. By devoting a section in this book to the subject, I hope to increase the popularity of taking photos instead of live animals.

The best advice is to keep it simple and lightweight. Most naturalists are already lugging around plenty of field equipment, and needlessly adding more is not desirable. In

Herpetologist Bill Leonard photographing Desert Spiny Lizards at Pyramid Lake in Northwestern Nevada. His equipment is ideal for wary reptiles when distance from the subject is required: an automated 35-mm SLR camera body, 200-mm macro lens, flash for fill light, and a sturdy tripod.

Reptiles rarely remain motionless for long. The latest high-tech camera equipment allows accurate efficiency to capture a fleeting image. A 100–300 zoom lens was used for this shot of a Western Fence Lizard and colorful Twining Snake Lily. The Nature Conservancy's Dye Creek Preserve, near Red Bluff, Sacramento Valley, northern California.

this regard, the conveniently sized 35 mm single-lens-reflex (SLR) camera with its seemingly endless array of interchangeable lenses, flashes, and other accessories is the ideal system for the nature photographer.

Technological advances in recent years have made the modern versions of these cameras a joy to use. Many readers probably already own older, totally manual 35 mm SLRs and these are fine for reptile photography. In fact, at least two non-electronic models are still available and are preferred by some photographers because "there are fewer complicated gizmos to go haywire." However, when buying a new one, I recommend the latest automated electronic cameras. I've used both types (and several brands) extensively during 30-plus years of photographing nature. The current computer-brained marvels boast lightening-fast autofocusing (most have lock-on tracking for moving subjects), astonishingly accurate "multipattern" metering, "smart flashes," power winders for rapid film advance and bursts of exposures, plus a host of other almost magical features. They have been very dependable and have allowed me to get shots of wildlife that were nearly impossible to achieve years ago. Free-roaming animals rarely stay still for prolonged periods in consistently good lighting. A just-captured snake is always intent on escape and will usually only briefly hold the pose you have spent an exasperating half hour coaxing it into. Quick efficiency and accuracy is a must or the shot will be missed completely, and there is frequently no second chance. Through automated complexity, nature photography is simplified. Instead of constantly making manual adjustments, we are largely freed to concentrate on composition and other creative considerations.

Actually, most of these cameras offer the ability to override all the automatic modes and be set on "M" (manual), should you choose to do so. In most situations I use a semi-automatic methodology by keeping my camera set on the aperture priority mode (you select the aperture; the camera automatically chooses the shutter speed.) This keeps me in manual control over depth-of-field when I adjust the *f*-stops, but simplifies things because the camera does the rest. Occasionally, if I think the camera's meter may have chosen the wrong exposure in a tricky lighting situation, I'll merely make a quick override adjustment with the exposure compensation dial.

The average amateur need not purchase a top-of-the-line camera body, which is usually quite expensive. It should be noted, though, that you generally get what you pay for. The professional-level models are more rugged and better sealed against dust and moisture. Nevertheless, all the major brand systems offer affordable, lower-level SLRs that are compactly lightweight and sport a wide range of high-tech features. I particularly recommend acquiring a camera body that provides a depth-of-field preview function, dial-in exposure compensation (preferably with a plus/minus scale conveniently visible in the viewfinder), power winder, built-in pop-up flash for occasionally needed fill light and, if possible, a mirror lock-up (rare in entry-level cameras) to reduce vibration when using slow shutter speeds.

Outdoor enthusiasts occasionally ask for my opinion about what to purchase for an uncomplicated nature photography outfit. My usual suggestion is a compact, electronic camera body (with the above-mentioned useful features) and two autofocus zoom lenses: one in the 28–100 mm or 35–80 mm range, and another of around 80–200 mm or 100–300 mm. These optics provide a useful selection of focal lengths from wide angle to moderate telephoto. For the herp photographer, a couple of other accessories will be required. One is an extension tube of around 25 mm to place between the camera and lens, which permits closer focusing when taking frame-filling pictures of small reptiles. A set of high-quality supplementary diopter lenses that screw on the end of your regular lenses and increase magnification is another option. The other accessory is a separate flash unit that can be used off the camera in conjunction with an electronically compatible extension cord of about 30 in (76 cm) in length. The small, pop-up flashes incorporated into camera bodies do not work for close-up shots. Being positioned on top of the camera, light is aimed straight ahead and too high above the subject, instead of where the lens is pointed. An independent flash can be positioned to angle downward toward the subject or any direction you choose. All this equipment can be carried into the field in a modestly sized photo fannypack with room to spare for film, lens cleaner, and a candy bar.

As with the newer electronic camera bodies, I highly recommend the latest breed of amazing high-tech flash units. Get one that is the same brand as your camera. The flash and camera will communicate with each other, automatically adjusting apertures and shutter speeds, while accurately balancing flash duration with ambient light that is metered through-the-lens (TTL) off the film plane. Be sure the flash has a dial-in plus/minus override adjustment so that flash duration can be easily fine-tuned to suit your taste. These features make working with flash a breeze, compared to manual flashes that require the user to make often-complicated calculations. When flash is used as the primary lighting for close-up photography, sharp pictures are assured because any movement is frozen, such as camera shake or a wriggling reptile. Additionally, there is sufficient illumination to use small apertures for maximum depth-of-field while having the flexibility of a hand-held camera. If low-level

Dramatic close-ups of venomous snakes require a telephoto lens for safety. I used a 300-mm lens with a 1.4 teleconverter (making it a 420-mm optic) for this portrait of a Prairie Rattlesnake. Cedar Springs Wildlife Area, northwestern Colorado.

Naturalist Alice Elshoff carefully prodding a Great Basin Rattlesnake into an artistic pose for the author. Big Basin, southeastern Oregon.

natural light is the primary illumination (perhaps with only a small amount of fill-flash), a slow shutter speed will be required. Consequently, you will need to mount your camera on a tripod to hold it steady and the reptile must be motionless. Otherwise, the resulting pictures will lack sharpness. Although most people dislike carrying a tripod, for razor-sharp photos it is frequently necessary (preferably equipped with a ball-head for ease of use).

If your budget permits, a macro lens is the best optical tool for taking close-up photographs. They are pricey but are always the most superbly sharp lenses available in any manufacturer's line of optics. Macro lenses generally come in approximately 50 mm, 100 mm, and 200 mm focal lengths, and most allow life-size (1:1) magnification without added accessories. If only one can be chosen, go with the 100 mm size. At 50 mm you are often too close to the reptile and might frighten it away, whereas the 200 mm macro keeps you too far removed from a captured reptile to quickly reach out and halt an attempted escape. The longer telephoto lenses are best reserved for distant, unrestrained reptiles that cannot be approached closely. One of my favorite outfits for this is a 300 mm lens with a 1.4 teleconverter that transforms it into a 420 mm optic. I use it for wary lizards, basking turtles, and venomous snakes.

Although this picture of a Red-spotted Garter Snake (a colorfully photogenic subspecies of the Common Garter Snake) looks natural, it was staged in the field. The snake was found crossing a road in harsh midday light. I moved it to the dead log, created shade with an umbrella, and used some fill-flash for softer light. Wilson Wildlife Area, near Corvallis, Willamette Valley, northwestern Oregon.

When choosing film, most nature photographers prefer the slower 50 to 100 speed slide films with their saturated colors, high contrast, and ultra-sharpness. Faster 400 to 800 speed films are usually grainy and less sharp, along with having reduced contrast and colors (although new versions keep improving). Print film generally compounds all these problems. Slides have the advantage of providing more options. Besides sharing your photographic work in an educational slide-show, prints can be made from them as well. Additionally, editors prefer slides for publication should you ever try to sell your creations. As more affordable 35 mm SLR digital camera bodies enter the market, the possibilities transcend traditional film.

Untouched, free reptiles can occasionally be photographed in great natural light while they strike an artistic pose in front of a pleasing background. However, more often than not, the light is harsh or drab, the background is visually unappealing, and the reptile hides as soon as you begin stalking it. In most cases, the best procedure is to "improve upon nature" by briefly capturing the specimen and placing it on a suitable background for a photo session. I always try to choose a spot that typifies that particular species' habitat but is not so cluttered that it partially conceals the animal or distracts the viewer's eye in the resulting photo. When photographing during glaringly bright, midday light, move into the shade of vegetation or block the sun with your body, hat, or an umbrella. Then you can create softer, pleasing light with a bit of subtle fill-flash or a reflector (use white paper or a sheet of crinkled aluminum foil to bounce diffused light onto the subject).

If the reptile is totally uncooperative and keeps moving about, place a snake bag or some other object over it. Feeling hidden, in a few moments it may calm down and remain properly posed when the cover is removed. Although best avoided whenever possible, a last resort tactic is to place the reptile in an ice chest to make it lethargic. However, this will cause some species of lizards to become darker or dull in coloring, and if the reptile is overly chilled it will appear lifeless in the photo. If you desire both the animal and its habitat to be completely in focus from the foreground to the background of the picture, use small apertures like *f*/22, or *f*/32. For a reptile posed against a simple background of out-of-focus, blurred colors, try a larger aperture like *f*/2.8 or *f*/4 that provides shallow focus. Use your camera's depth-of-field preview function to visually check the effect before tripping the shutter.

It is really personal preference, but when a flash is the primary source of illumination, the lighting in the picture appears harsh and unnatural to me. Another disadvantage arises when the rear habitat is located some distance away from the reptile. There will be flash illumination fall-off, resulting in the foreground animal being properly lit, but the background will be black in the final picture. If the reptile will remain immobile, I remedy these flash-related problems by setting up on a tripod and using a slower shutter speed, allowing natural light to dominate the exposure. I then add only a hint of fill-flash to give a little zip to the shot and some sparkle in the creature's eyes. With this method the background does not end up as an inky shadow in the photograph. Flash photography offers many options. Some photographers attach a special bracket to their camera that holds a flash unit and can be adjusted to several configurations, both above or beside the camera body. I prefer to merely use my hand to hold the flash exactly where I want it. In my experience, it is distracting and wastes time to keep readjusting a bracket-held flash when I'm rapidly changing shooting positions. I find the hand-held method to be efficiently simpler and there is one less piece of equipment to carry. In contrast, friend and fellow nature photographer Terry Steele often uses three flash units mounted on small supportive stands for herp photographs. He positions them for a combination of cross-lighting that softens shadows and illuminates the background habitat with beautiful results. Find what works best for you.

Using a large aperture of f/4 with my 300-mm telephoto lens produced a shallow zone of focus and caused the background to become a pleasant blur of colors. Western Fence Lizard, Cove Palisades State Park, central Oregon.

During the spring breeding season, male Great Basin Collared Lizards become very territorial and can often be closely approached. I used a 100-mm macro lens while reclining on the ground to achieve this low-angle view, making the reptile look like a giant dinosaur. Pueblo Mountains, southeastern Oregon.

Turtle Morphology & Quick Key

Scutes on the shell of a turtle

1. Marginals
2. Nuchal
3. Vertebrals
4. Costals
5. Gulars
6. Humerals
7. Pectorals
8. Abdominals
9. Femorals
10. Anals

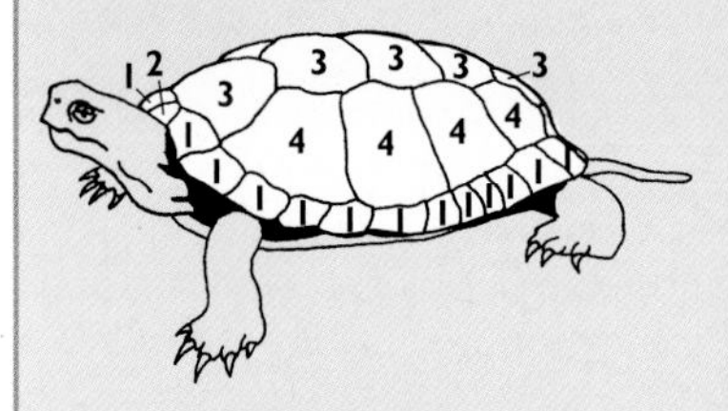

Carapace

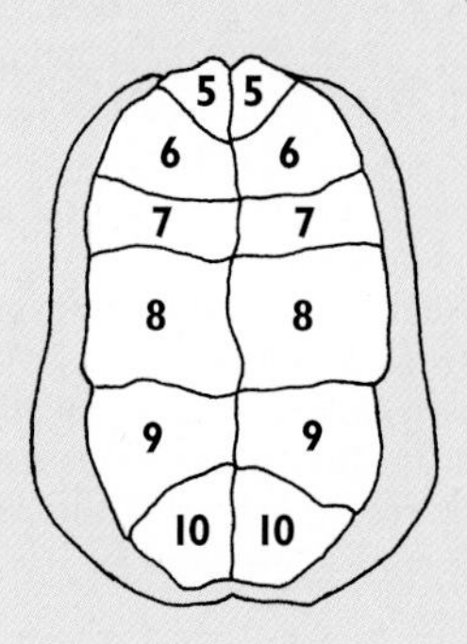

Plastron

Western Pond Turtle
(*Clemmys marmorata*)..........................Page 72

Neck and legs drab brown, sometimes with cream to yellowish-white tinges. Plastron pale cream to yellowish white with large, dark brown markings.

Painted Turtle
(*Chrysemys picta*)...............................Page 76

Neck and legs marked with yellow stripes. Plastron bright red with large, dark marking in center.

Slider
(*Trachemys scripta*).......................... Page 250

Neck and legs marked with yellow stripes (sometimes faint or absent on old males). May have patch of red on side of head. Plastron yellowish with large, dark blotches.

Snapping Turtle
(*Chelydra serpentina*)....................... Page 250

Head and legs massive. Tail thick and long (nearly ⅔ length of shell). Carapace often has protruding, knobby keels. Plastron extremely small and narrow.

Lizard Morphology & Quick Key Finder

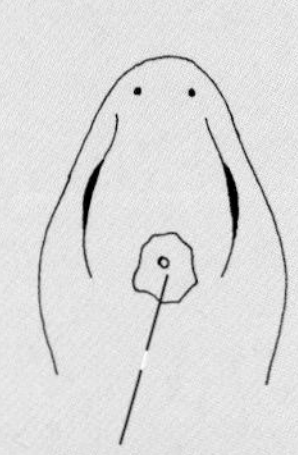

Parapineal organ (parietal "eye") in the center of the interparietal scale on an iguanid (phrynosomatid) lizard's head

Gular fold

Vent region of an iguanid lizard

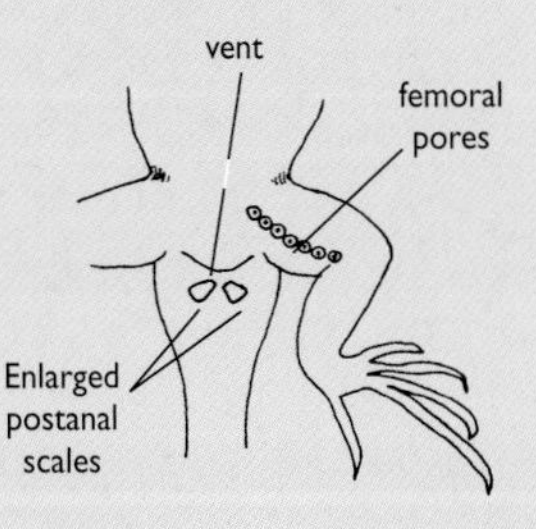

Lizard scale types

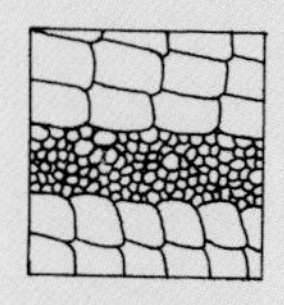

Alligator lizard's skin fold along side of body with granular scales within

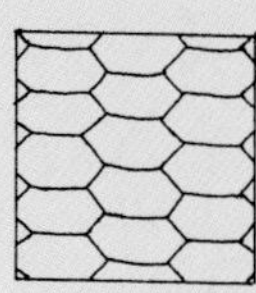

Smooth

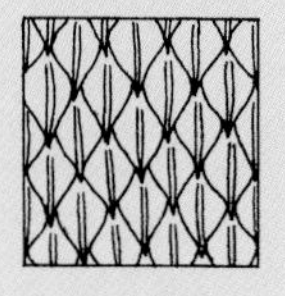

Pointed

Granular

Lizards with Granular Scales

.... Page 58

All or nearly all dorsal and ventral scales small and rounded and do not overlap.

Lizards with Pointed Scales

.... Page 59

Dorsal scales pointed, giving a prickly appearance. (On one species, pointed scales are small and have only slightly rough texture.) Males of all species in group have patches of blue on bellies and throats. Blue patches often faint or absent on females.

Horned Lizards .... Page 60

Enlarged spines ("horns") at rear of head (reduced to small, pointed nubbins on one species). Scattering of smaller spines on back. Fringe of small spines along each side of rounded, flattened body.

Striped Lizards .... Page 61

Pattern of lengthwise stripes on body. Tail blue on juveniles and young adults.

Alligator Lizards .... Page 62

Large, squarish dorsal and ventral scales arranged in lengthwise rows. Body elongated, with short legs and large head. Distinctive fold of skin along both sides of body that have small granular scales within.

Lizards with Granular Scales

Zebra-tailed Lizard

(*Callisaurus draconoides*)........................Page 82

Underside of tail boldly banded with black and white. In our region, confined to northwestern Nevada.

Long-nosed Leopard Lizard

(*Gambelia wislizenii*)................................Page 90

Distinctive dorsal pattern of dark spots. No black-and-white banding on underside of tail.

Great Basin Collared Lizard

(*Crotaphytus bicinctores*)........................Page 86

Distinctive black-and-white markings on neck. Large head, narrow neck, bulky body, large hind legs, and long, thick tail.

Common Side-blotched Lizard

(*Uta stansburiana*)................................Page 110

Black blotch on each side of body behind front legs. Gular fold at rear of throat.

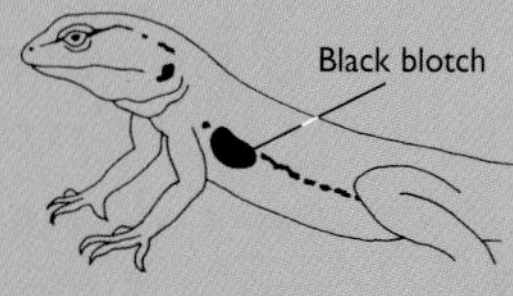

Ornate Tree Lizard

(*Urosaurus ornatus*)..............................Page 114

All dorsal scales granular except for narrow strip of enlarged scales running down middle of back. Fold of skin along each side of body and gular fold at rear of throat. Males have blue patches on belly. In our region, confined to northeastern Utah, northwestern Colorado, and extreme southwestern Wyoming.

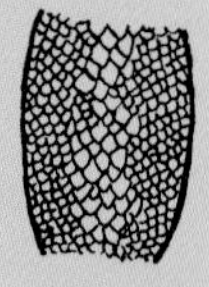

Narrow strip of enlarged scales down mid-back

Lizards with Pointed Scales

Western Fence Lizard

(Sceloporus occidentalis) ....Page 98

Dorsal scales moderately large and noticeably prickly. Large, solid blue patch on throat (sometimes divided into 2 sections; faint or absent on females).

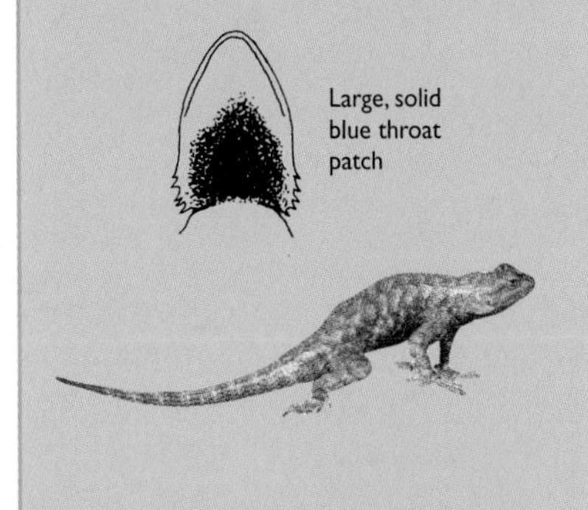

Desert Spiny Lizard

(Sceloporus magister) ....Page 94

Dorsal scales very large and extremely prickly. All scales have sheen. Large, solid blue patch on throat (faint or absent on females). Black, wedge-shaped mark on each side of neck. In our region, confined to northwestern Nevada.

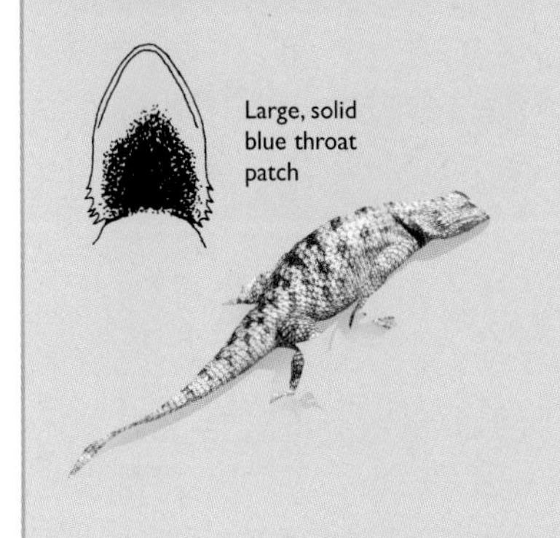

Eastern Fence Lizard

(Sceloporus undulatus) ....Page 102

Dorsal scales moderately large and noticeably prickly. Small patch of solid blue at each rear corner of throat (faint or absent on females). In our region, confined to northeastern Utah, northwestern Colorado, and extreme southwestern Wyoming.

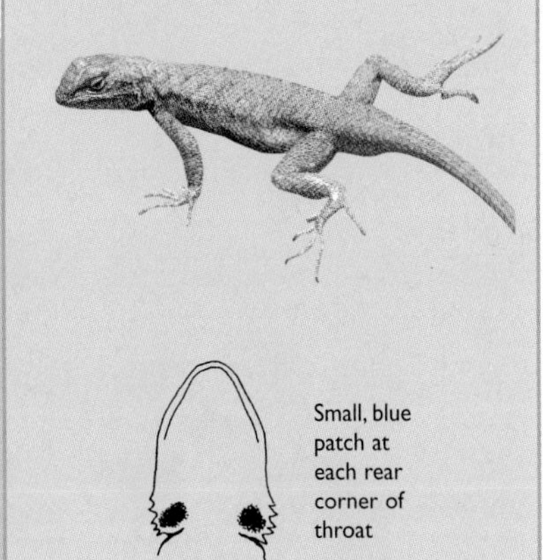

Sagebrush Lizard

(Sceloporus graciosus) ....Page 106

Dorsal scales small and only slightly rough. Mottled blue pattern on throat (often absent on females).

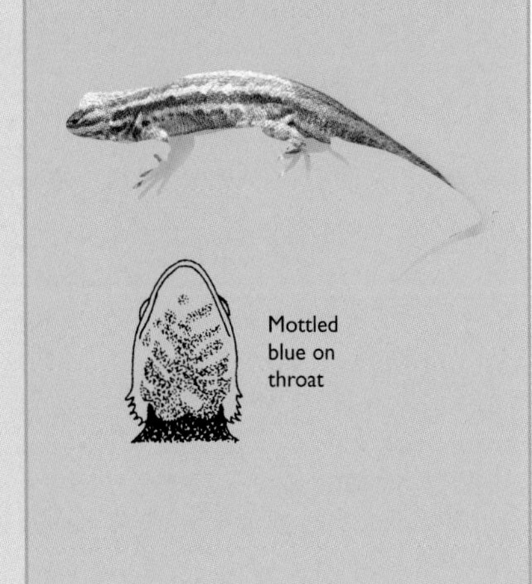

Horned Lizards

Desert Horned Lizard

(*Phrynosoma platyrhinos*)......................Page 118

Crown-like, large spines ("horns") project horizontally from rear of head.

Crown-like, large horns

Greater Short-horned Lizard

(*Phrynosoma hernandesi*)......................Page 126

Stubby horns project nearly horizontally from rear of head. Noticeably wide, deep notch separates horns left and right at rear of head, giving heart-shaped appearance to head when viewed from above.

Stubby horns on heart-shaped head

Pigmy Short-horned Lizard

(*Phrynosoma douglasi*)..........................Page 122

Horns merely small, pointed nubbins that project almost vertically from rear of head. Lacks heart-shaped, deeply rear-notched head. Rarely exceeds 3½ in (9 cm) in total length.

Horns merely small, pointed nubbins

Striped Lizards

Western Skink

(*Eumeces skiltonianus*) .... Page 130

Entire body covered with shiny, smooth, rounded scales. Juveniles and younger adults have brilliantly blue tails.

Plateau Striped Whiptail

(*Cnemidophorus velox*)........................... Page 251

Small, granular scales on upper parts and sides of body. Large, squarish scales on belly and chest. Gular fold on throat. Vivid, even-edged dorsal stripes continue along entire length of back. Belly uniform white without dark spotting. Juveniles have blue tails. Confined to vicinity of Cove Palisades State Park, Jefferson County, central Oregon.

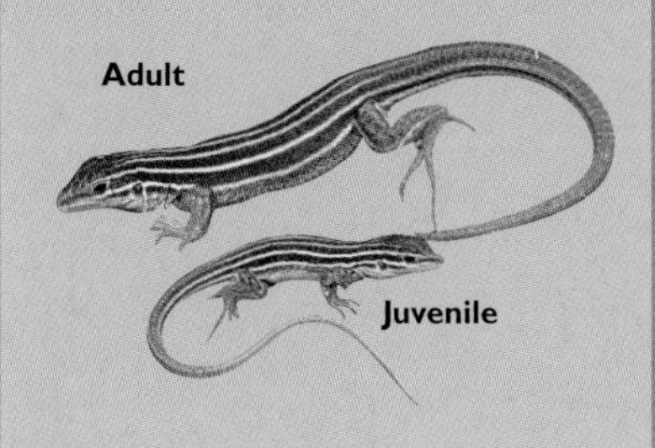

Western Whiptail

(*Cnemidophorus tigris*) .... Page 134

Small, granular scales on upper parts and sides of body. Large, squarish scales on belly and chest. Gular fold on throat. Dorsal stripes on adults often dull and become faded and indistinct toward rear of body. Belly gray to bluish gray with dark spots. Juveniles have vivid dorsal stripes and blue to bluish-gray tails.

Adult

Juvenile

Alligator Lizards

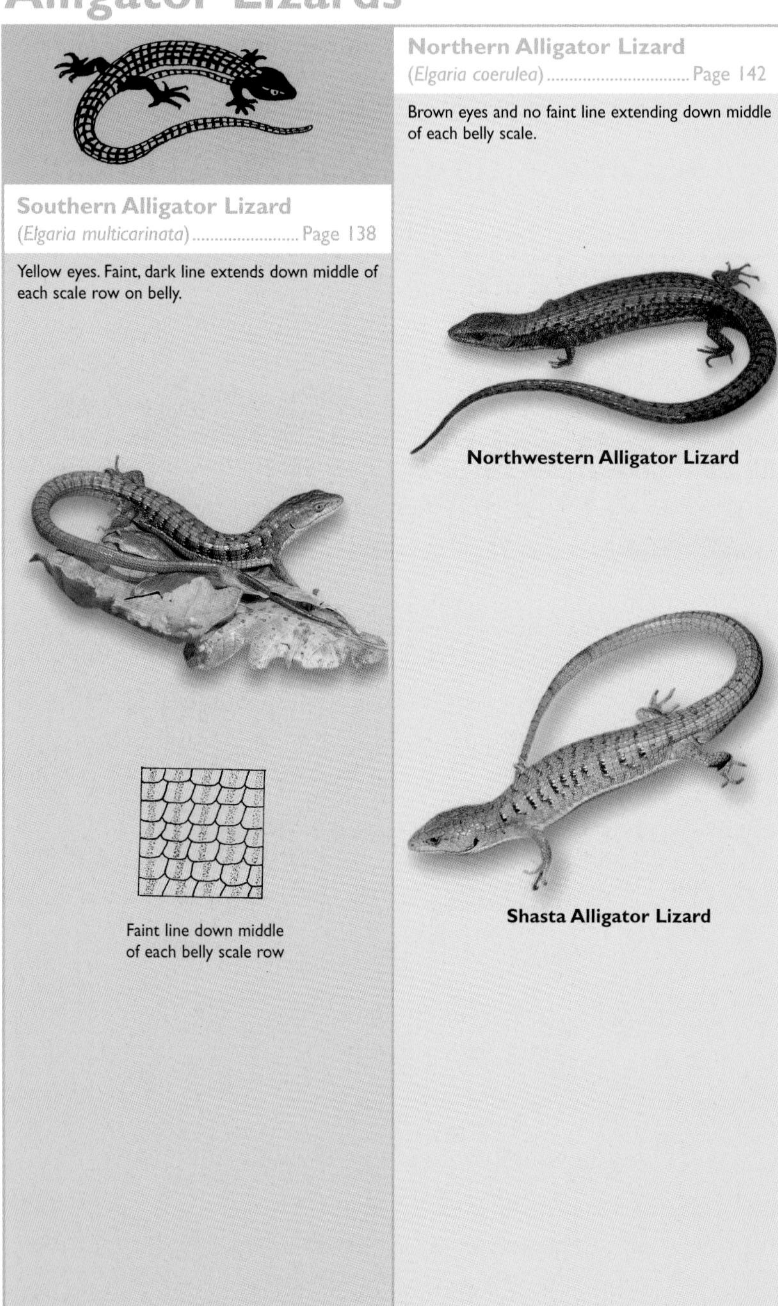

Southern Alligator Lizard

(*Elgaria multicarinata*) .... Page 138

Yellow eyes. Faint, dark line extends down middle of each scale row on belly.

Faint line down middle of each belly scale row

Northern Alligator Lizard

(*Elgaria coerulea*) .... Page 142

Brown eyes and no faint line extending down middle of each belly scale.

Northwestern Alligator Lizard

Shasta Alligator Lizard

Snake Morphology & Quick Key Finder

Major head scales on colubrid snakes

1. Rostral
2. Upper labials
3. Lower labials
4. Mental
5. Internasals
6. Prefrontals
7. Frontal
8. Supraoculars
9. Parietals

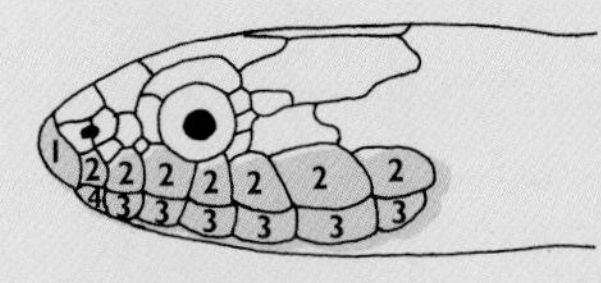

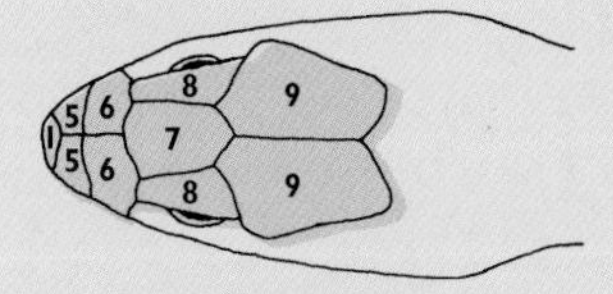

Snake scale types

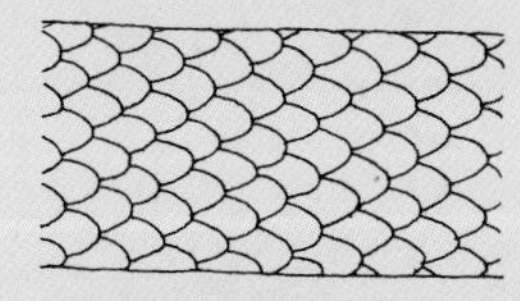

Smooth

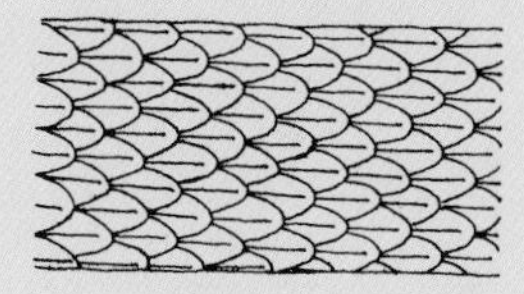

Keeled

Vent region of a snake

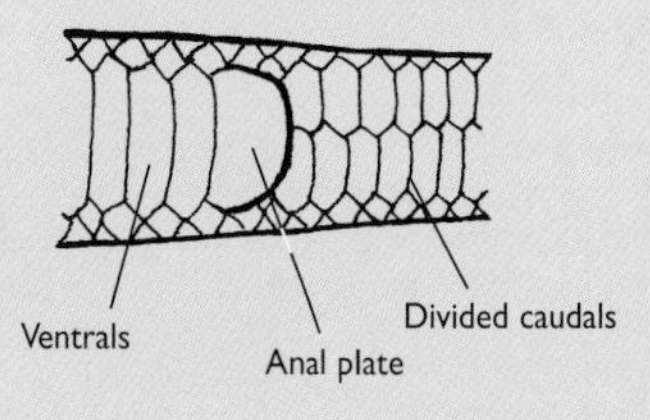

Uniformly Colored Snakes .... Page 64

No definite overall pattern. Dorsal coloring of body tan, brown, reddish brown, gray, olive, or green. May be pattern on neck area of dark crossbars or band of orange or yellow.

Blotched Snakes .... Page 65

Pattern of blotches or large spots. Some species may have narrow dorsal blotches that extend downward onto sides of body to form crossbars (may fade away toward tail).

Crossbanded Snakes .... Page 66

Bold dorsal pattern of alternating light and dark bands. Pattern may be bicolored (black and white; brown and white; or reddish orange and black), or tricolored (red, black, and white.)

Striped Snakes

with smooth scales .... Page 67
with keeled scales (garter snakes) .... Page 68

Dorsal pattern of light, lengthwise stripes along mid-back and/or sides of body. On some species, stripes will be most evident in area of neck, fading away toward mid-body.

Uniformly Colored Snakes

Rubber Boa

(*Charina bottae*) .... Page 148

Tail short and blunt. Thick-bodied. Soft, loose skin with small, smooth dorsal scales. Small eyes with vertical pupils.

Racer, adult

(*Coluber constrictor*) .... Page 162

Tail long and narrow. Slim-bodied. Large, smooth dorsal scales. Large eyes with round pupils.

Coachwhip, adult

(*Masticophis flagellum*) .... Page 174

Tail long and narrow. Slim-bodied. Large, smooth dorsal scales. Large eyes with round pupils. Dark brown to reddish-brown, irregular crossbars on neck that become progressively more faint toward mid-body. Dark edging on dorsal scales of rear parts of body and tail. In our region, confined to northwestern Nevada.

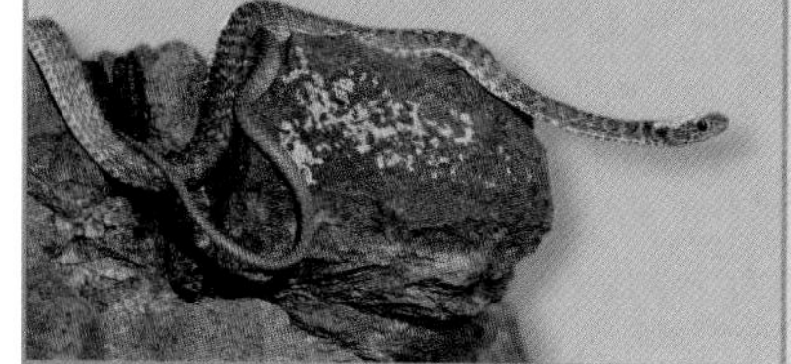

Sharp-tailed Snake

(*Contia tenuis*) .... Page 152

Tail ends abruptly with sharp spine. Round pupils. Black crossbar along upper half of each ventral scale on belly creates alternating, ladder-like pattern. Faint coppery-red line along each side of back (juveniles often have solid red or orangish-red backs).

Black crossbars on belly

Ring-necked Snake

(*Diadophis punctatus*) .... Page 156

Reddish-orange or yellowish-orange ring on neck (often absent in Utah and southeastern Idaho populations). Belly bright reddish orange, yellowish orange, or yellow and sharply set off from uniform slate-gray dorsal color.

Smooth Green Snake

(*Opheodrys vernalis*) .... Page 160

Dorsal coloration bright grass green. Nostrils located in center of a single scale on each side of head. In our region, confined to northern Utah, northwestern Colorado, and south-central Wyoming.

Ground Snake, uniform morph

(*Sonora semiannulata*) .... Page 236

This morph lacks dorsal markings and is uniformly tan or olive brown. Dark spot on each dorsal scale gives dappled appearance (most apparent on sides of body). Head short and barely wider than neck. In our region, confined to northwestern Nevada, southeastern Oregon, and southwestern Idaho.

Blotched Snakes

Gopher Snake

(Pituophis catenifer)................................Page 182

Strongly keeled scales on back, grading to smooth scales on sides of body. Round pupils.

Corn Snake

(Elaphe guttata)....................................Page 186

Weakly keeled scales on back, grading to smooth scales on sides of body. Distinctive, V-shaped marking on top of head. Round pupils. In our region, confined to northeastern Utah.

Racer, juvenile

(Coluber constrictor)..............................Page 162

Smooth dorsal scales. Round pupils. Blotches confined to back and do not extend downward onto sides of body. Markings most vivid on neck, fading away completely toward tail.

Coachwhip, juvenile

(Masticophis flagellum).........................Page 174

Smooth dorsal scales. Round pupils. Nearly entire length of snake marked with blotches that extend downward onto sides of body to form irregular crossbars. Blotches on neck usually black. In our region, confined to northwestern Nevada.

Night Snake

(Hypsiglena torquata)...........................Page 240

Smooth dorsal scales. Vertical pupils. Distinctive, dark, collar-like marking on neck.

Western Rattlesnake

(Crotalus viridis).....................................Page 244

Strongly keeled dorsal scales. Vertical pupils. Horny rattle or nubbin ("button") on end of tail.

Crossbanded Snakes

California Mountain Kingsnake

(*Lampropeltis zonata*)......................... Page 194

Alternating red, black, and white crossbands that encircle body. White bands usually of uniform width (neither widen nor narrow on lower sides of body). Snout black. In our region, confined to northern California, southwestern Oregon, and south-central Washington.

Common Kingsnake

(*Lampropeltis getula*).......................... Page 190

Alternating black and white crossbands that encircle body (sometimes dark brown and white in Sacramento River drainage). No red in pattern. In our region, confined to northern California, southwestern Oregon, and northwestern Nevada.

Ground Snake, crossbanded morphs

(*Sonora semiannulata*)........................ Page 236

Some morphs have alternating reddish-orange and black crossbands that do not encircle body. Less commonly, may have crossbands of black and white or black and gray. Belly uniform white or pale yellow. Dark spot on each dorsal scale gives dappled appearance (most apparent on sides of body). Head short and barely wider than neck. In our region, confined to northwestern Nevada, southeastern Oregon, and southwestern Idaho.

Sonoran Mountain Kingsnake

(*Lampropeltis pyromelana*) .... Page 198

Alternating red, black, and white crossbands that encircle body. White bands often become narrower on lower sides of body. More than 43 white bands in total on body and tail. Snout usually completely white. Neck thin. Head prominent and somewhat flattened. In our region, confined to northwestern Utah.

Milk Snake

(*Lampropeltis triangulum*).................. Page 202

Alternating red, black, and white crossbands that usually encircle body. White bands become wider on lower sides of body. Less than 38 white bands in total on body and tail. Snout usually black (sometimes with flecks and mottlings of white or red). Neck short and thick. In our region, confined to northern Utah and northwestern Colorado.

Stansbury Mountains variation

Long-nosed Snake

(*Rhinocheilus lecontei*) .... Page 206

Alternating orangish-red, black, and white crossbands that do not encircle body. Pattern distinctively speckled (especially on sides of body). Belly uniform white or cream. Most caudal scales on underside of tail arranged in single, undivided row. Snout elongated and pointed, with enlarged rostral scale. In our region, confined to northwestern Nevada, northeastern California (Honey Lake Basin), northwestern Utah, and southwestern Idaho.

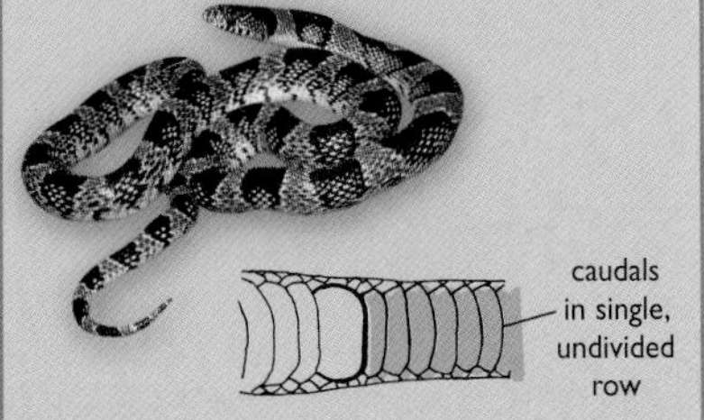

Striped Snakes with Smooth Scales

Striped Whipsnake

(*Masticophis taeniatus*)......................... Page 166

Light stripe along each side of back, with no mid-dorsal stripe. Narrow black line (often broken into dashes) runs down middle of each stripe. Underside of tail coral pink. Scales on head distinctively edged with white.

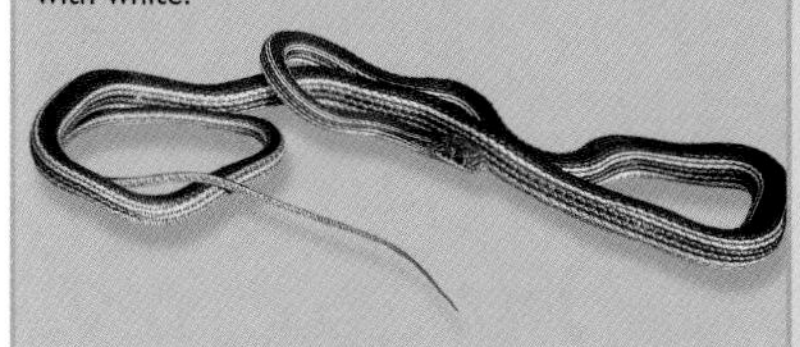

California Whipsnake

(*Masticophis lateralis*)............................ Page 170

Like Striped Whipsnake, but lacks black line down middle of each light stripe, and head scales are not edged with white. In our region, confined to Sacramento River drainage.

Western Patch-nosed Snake

(*Salvadora hexalepis*)............................ Page 178

Broad (3 scale rows wide), light mid-dorsal stripe. Large, triangular-shaped rostral scale that curves back over top of snout. In our region, confined to northwestern Nevada and Honey Lake Basin of northeastern California.

Ground Snake, striped morph

(*Sonora semiannulata*) .... Page 236

This morph has narrow to wide, orange mid-dorsal stripe. Dark spot on each dorsal scale gives dappled appearance (most apparent on sides of body). Head short and barely wider than neck. In our region, confined to northwestern Nevada, southeastern Oregon, and southwestern Idaho.

Striped Snakes with Keeled Scales

A: Garter Snakes with Seven Upper Labials

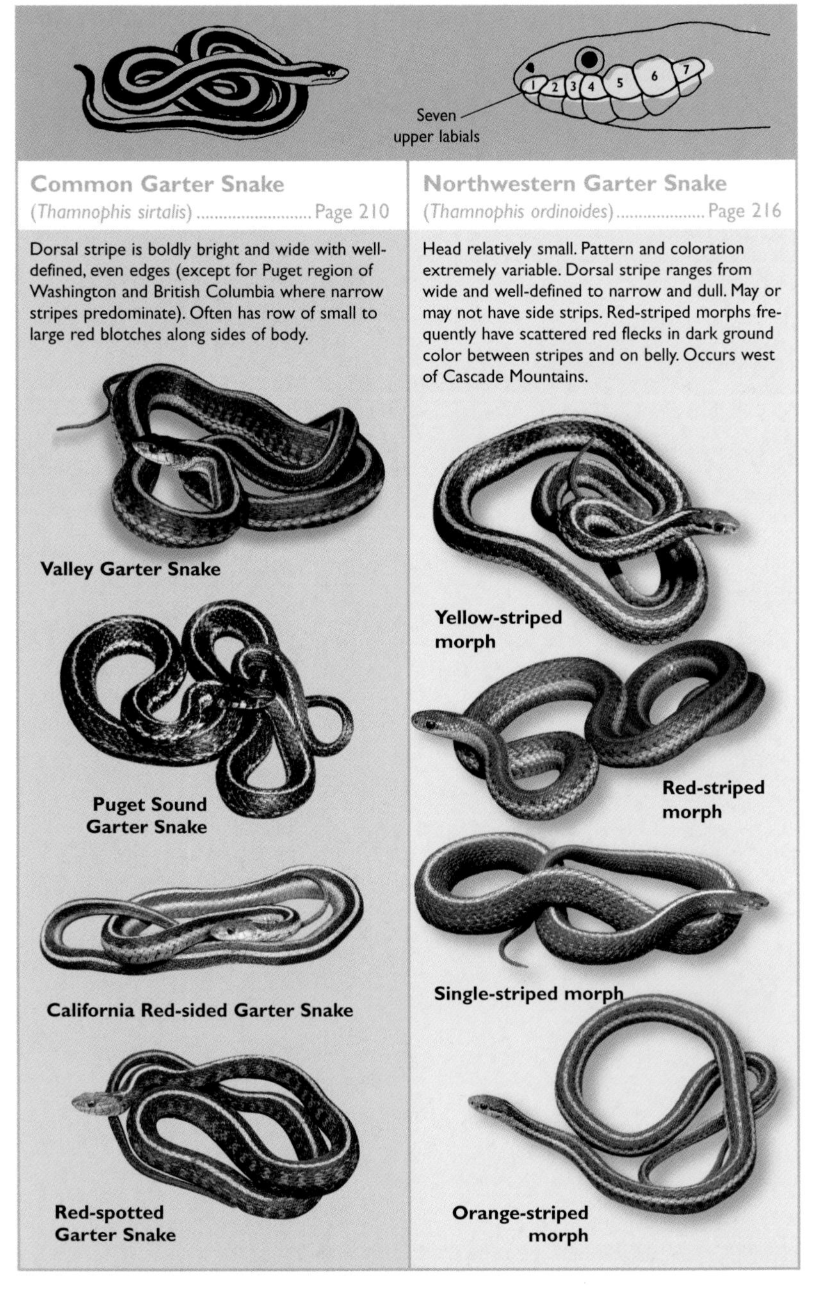

Common Garter Snake

(*Thamnophis sirtalis*) .... Page 210

Dorsal stripe is boldly bright and wide with well-defined, even edges (except for Puget region of Washington and British Columbia where narrow stripes predominate). Often has row of small to large red blotches along sides of body.

Northwestern Garter Snake

(*Thamnophis ordinoides*) .... Page 216

Head relatively small. Pattern and coloration extremely variable. Dorsal stripe ranges from wide and well-defined to narrow and dull. May or may not have side strips. Red-striped morphs frequently have scattered red flecks in dark ground color between stripes and on belly. Occurs west of Cascade Mountains.

Striped Snakes with Keeled Scales

B: Garter Snakes with Eight Upper Labials

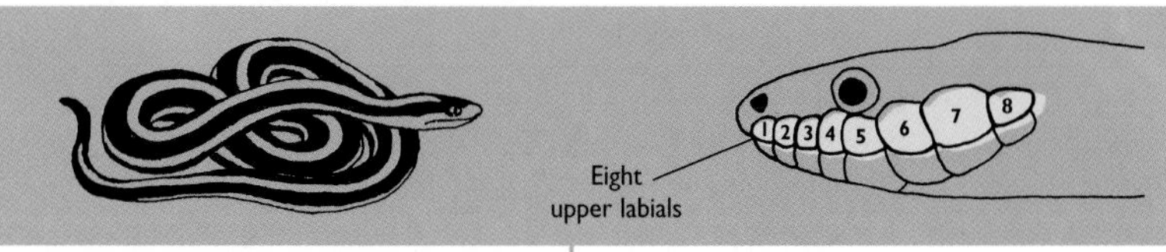

Pacific Coast Aquatic Garter Snake

(*Thamnophis atratus*) .... Page 226

Internasal scales on top of snout longer than wide and pointed at front end. Two color/pattern variations: spotted morph and striped morph. Underside of tail salmon orange or pinkish purple. In our region, confined to southwestern Oregon and northern California.

Spotted morph

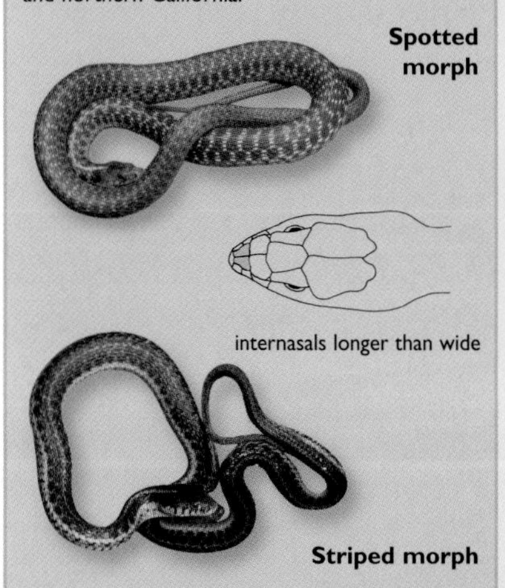

internasals longer than wide

Striped morph

Sierra Garter Snake

(*Thamnophis couchii*)........................... Page 232

Internasal scales on top of snout longer than wide and pointed at front end. Head narrow with distinctively pointed snout. Underside of tail often nearly black with no salmon-orange or pinkish-purple coloration. In our region, confined to Sierra Nevada.

internasals longer than wide

Sierra Garter Snake

Western Terrestrial Garter Snake

(*Thamnophis elegans*) .... Page 220

Internasal scales on top of snout wider than long and not pointed at front end. Coloration and pattern varies considerably according to geographic location. Darker melanistic individuals are common.

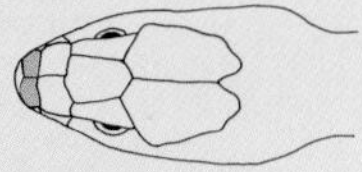

internasals wider than long

Coast Garter Snake

Mountain Garter Snake

Willamette Valley variation

Wandering Garter Snake

darker Puget Sound variation

light variation

The Turtles

Duckweed-coated Painted Turtle. National Bison Range, western Montana.

Western Pond Turtle

Clemmys marmorata

Adult. Shasta Valley, northern California.

IDENTIFICATION: Adults are 4 in (10 cm) to nearly 9 in (23 cm) in shell length; hatchlings are approximately 1 in (2.5 cm) in shell length. This turtle is sometimes called "mud turtle" because **its carapace is a rather drab dark brown, olive brown, or almost black**. There is usually a pattern of dark flecks and fine mottling on the shell, although it may be faint. **Below, the plastron is pale cream to yellowish white, with large dark brown markings**. Some individuals from aquatic habitats that contain tannin may have completely dark brown staining on the plastron. The head and legs usually show some cream to yellowish-white coloring as well. Males typically have unmarked, light throats, low-domed carapaces, and concave plastrons. Females generally have dark markings on the throat, high-domed carapaces, and convex plastrons.

VARIATION: Although there is considerable variability in the extent and pattern of the shell's dark flecks and mottlings throughout this turtle's range, only two subspecies have been described. The race found in our region is the NORTHWESTERN POND TURTLE (*C. m. marmorata*). Some authorities have noted that populations in the Columbia River Gorge differ in certain respects.

SIMILAR SPECIES: The native Painted Turtle (p. 76) is readily distinguished by the yellow stripes on its legs and neck and the bright red markings on its plastron. Two non-native species, the Slider (p. 250) and the Snapping Turtle (p. 250), have been introduced into several parts of the Northwest. Like the Painted Turtle, the Slider is easily differentiated from the Western Pond Turtle by the yellow striping on its legs and neck (although it is sometimes faint or

absent in old males). The Snapping Turtle has a drab-brown shell coloration similar to the Western Pond Turtle, but it differs greatly in form: the head and legs are comparatively massive; the thick tail is often nearly two-thirds the length of the shell; the carapace often has protruding, knobby keels; and the plastron is extremely small and narrow.

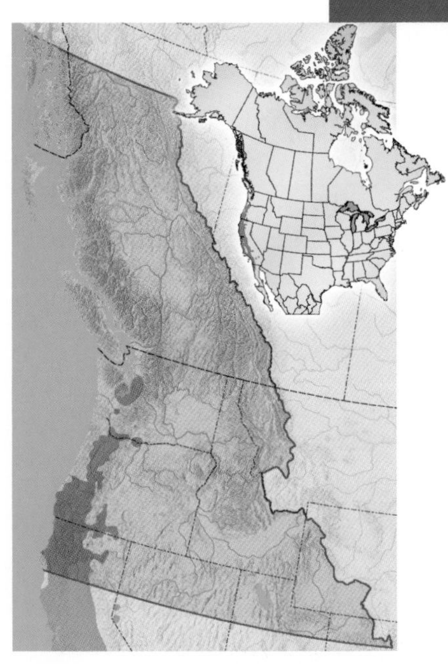

DISTRIBUTION: Although not as abundant in our region as it was in the earlier part of the 20th century, this turtle is still found at scattered localities throughout northern California, and the Rogue, Umpqua, and Willamette drainages of western Oregon. Isolated populations occur on both the Oregon and Washington sides of the Columbia River Gorge. This species ranges east of the Cascades along the Pit River in California and into the Klamath Basin of south-central Oregon and north-central California via the Klamath River. The Western Pond Turtle formerly occurred in western Washington, but it now appears that natural populations no longer exist there. Most individuals occasionally found in the Puget Lowlands are probably introductions. Old records from the 1930s for the Vancouver area of British Columbia are also probably based upon transplanted turtles from elsewhere. Currently, there are introduced individuals in Oregon along the Deschutes River at Bend and in the John Day River drainage near Canyon City. The Western Pond Turtle occurs at elevations from just above sea level to slightly over 4,000 ft (1,220 m) in the Northwest.

HABITAT AND BEHAVIOR: This reptile is most common in the mud-bottomed ponds, lakes, sloughs, marshes, and slow-moving rivers of valleys. The Western Pond Turtle is also sometimes found along the quiet pools of faster-flowing rocky tributaries in the adjacent foothills and mountains. Basking sites for thermoregulation are an important component of its habitat. Look for individuals or groups of turtles sunning on floating logs, protruding rocks, exposed cattail mats, and open banks. Use stealth and binoculars, because these shy creatures have excellent eyesight and will often disappear into the water when a stalker is as far away as 100 yd (91 m). Usually, though, if an observer stays patiently hidden, the turtle will soon reappear. Foods for this species include crayfish, insects, amphibian eggs and larvae, and aquatic plants. Frogs and fish are occasionally eaten,

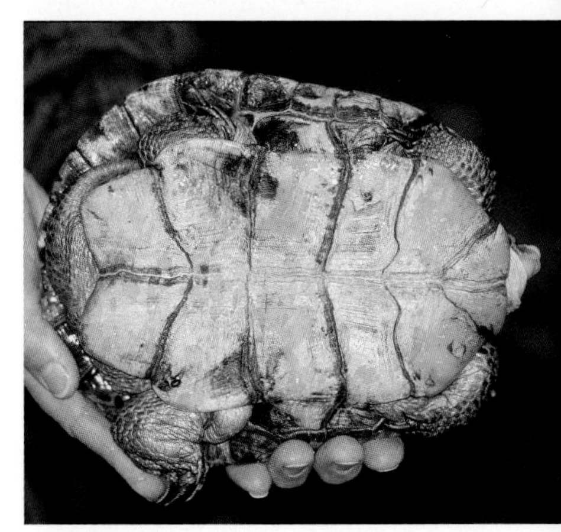

Plastron of adult. Puget Sound area, western Washington.

Western Pond Turtle

Adult. Near Corvallis, Willamette Valley, northwestern Oregon.

Basking adults. Columbia River Gorge, Washington.

but are probably scavenged as carrion in most instances. Nesting takes place from May to August, with present data indicating that some females lay eggs in alternate years, while others produce two clutches annually. A nest site with ample sunshine is chosen (usually a south- or west-facing slope) and may be as far as 1/4 mi (400 m) or more from water. After digging a nest hole, the female deposits and buries 1 to 13 eggs. Hatching takes place in 70 to 110 days, with most hatchlings overwintering in the nest at northerly latitudes. The lifespan of this long-lived turtle sometimes exceeds 50 years. Although the Western Pond Turtle is still moderately common in many parts of northern California and southwestern Oregon, only small, fragmented breeding populations exist in most parts of the Willamette Valley north of Lane County. Its decline is due primarily to habitat loss, the destruction of nesting sites by agricultural practices, and predation on juvenile turtles by introduced bass and bullfrogs, and possibly to diseases acquired from non-native

turtles. Another contributing factor appears to be the abundance of raccoons at the edges of urban areas and around campgrounds where trash cans provide ample food. Enlarged populations of these opportunistic mammals results in an increase in their predation on turtle eggs.

Hatchling. Columbia River Gorge, Washington.

FIELD NOTES: 14 June 1998, 6:00 p.m. Janet Wagener, a local U. S. Forest Service employee and friend, was guiding me through some winding backroads in the Shasta Valley of northern California. It was a pleasant, sunny evening drive and our attention was divided between watching for crossing snakes on the blacktop and the spectacular views of towering, snow-clad Mt. Shasta on the southern horizon. Suddenly, our eyes were riveted back to the road where something was moving. Expecting a snake, it took a moment to register upon my mind that it was actually a Western Pond Turtle. Picking it up for closer examination, we immediately noticed that a portion of the turtle's rear shell had been crushed in the past and had healed. However, owing to the missing edge of the carapace, the tail and left rear leg were exposed and could not be drawn back into the shell for protection against predators. Once again, I was amazed at the resilience of reptiles. The turtle was a female and probably had been on its way from a nearby wetland to look for a nest site to lay eggs. It had survived personal injury and was still attempting to reproduce. After taking photos, we released her off the side of the road and hoped this lone individual would have success on her journey.

Adult female with a damaged shell. Shasta Valley, northern California.

Painted Turtle

Chrysemys picta

Adult. Captive, Brad's World of Reptiles, Corvallis, Oregon.

IDENTIFICATION: Adults range from 4 in (10 cm) to nearly 10 in (25 cm) in shell length; hatchlings are approximately 1 in (3 cm) in shell length. **The carapace is olive green to dark greenish black, with a contrastingly bright scarlet red plastron (sometimes faded pinkish orange)**. There is often a large, dark marking in the center of the plastron. The overall form of the shell is flat and wide. **The legs and neck are distinctively marked with yellow stripes**, and the claws of the front feet are long. Males tend to be significantly smaller and have comparatively longer front claws, and the plastron is more concave toward the rear than in females.

VARIATION: Of the four subspecies that are recognized across this turtle's coast-to-coast range in North America, only the Western Painted Turtle (*C. p. bellii*) occurs in the Northwest.

SIMILAR SPECIES: The Painted Turtle is easily differentiated from our other native species, the Western Pond Turtle (p. 72), by the bright red markings on the plastron and yellow stripes on the legs and neck. The Slider (p. 250), an eastern turtle that has been introduced into a number of areas in the Northwest, has similar yellow stripes on the neck and legs, but it lacks the red plastron. The Slider also differs

in often having a red "ear" stripe behind the eye.

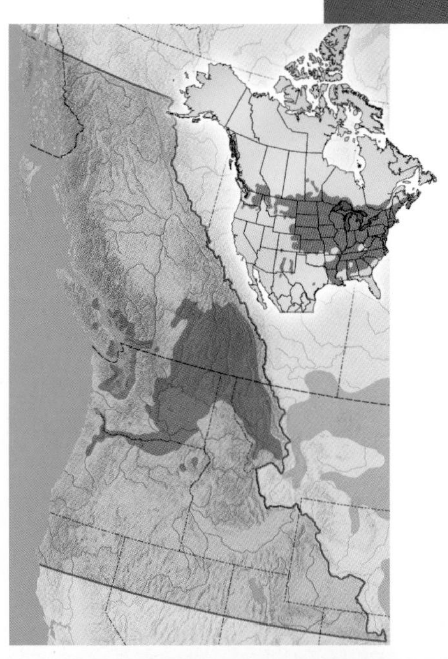

DISTRIBUTION: The Painted Turtle's primary range in the Northwest is east of the Cascade Mountains in the Columbia River drainage, usually below 3,500 ft (1,070 m). It is found throughout much of eastern Washington, the north-central and northeastern portions of Oregon, interior southern British Columbia, extreme northern Idaho, and western Montana. This species penetrates into northwestern Oregon and southwestern Washington via the Columbia River Gorge and up Oregon's Willamette River, at least as far south as the Salem area. Small, scattered populations occur in the Puget Sound vicinity of northwestern Washington and coastal southwestern British Columbia. It also occurs on Vancouver Island and some of the neighboring smaller islands, but these populations may be from introductions. The

Plastron of adult. Captive, Brad's World of Reptiles, Corvallis, Oregon.

Painted Turtle

Adult. Captive, Brad's World of Reptiles, Corvallis, Oregon.

Painted Turtle has been transplanted to many locations outside its natural range in the Northwest.

Basking adults. National Bison Range, western Montana.

HABITAT AND BEHAVIOR: Unlike the Western Pond Turtle, the Painted Turtle is usually not found in the faster-flowing, rocky tributaries of drainages. It prefers the bottomlands that provide sluggish rivers with oxbows, sloughs, ponds, lakes, and marshes. Ideal habitat includes relatively shallow aquatic areas with a muddy bottom and abundant plant growth, along with logs, exposed cattail mats, and open banks for basking. Recent studies in the Portland area of Oregon indicate that aquatic plant material makes up the majority of the diet. Additionally, a variety of animal foods may be eaten, such as insects, crayfish, earthworms, amphibian larvae, frogs, small fish, and carrion. Females deposit a yearly clutch of 1 to 20 eggs during June or July. Moving as far as nearly 500 ft (150 m) from water, the female excavates a

nest hole in soil that is exposed to sunshine. Records indicate that the incubation time ranges from 72 to 104 days. Eggs and hatchlings may sometimes overwinter in the nest.

Hatchling. Captive, Brad's World of Reptiles, Corvallis, Oregon.

FIELD NOTES: 11 August 1998, 5:30 p.m. While exploring the perimeter of a large pond at the National Bison Range in western Montana, I spotted a group of Painted Turtles sunning on a floating log near the water's edge. I wanted pictures of this photogenic assemblage, but I knew from past experience that it would require time and patience to accomplish. Far too often, such attempts are met with failure unless a blind is used. Even the slightest movement is usually noticed by these wary creatures, causing them to dive into the water. Luckily, there was a slope leading up to the side of the pond where the turtle log was located. Clutching my tripod-mounted camera, I bent over to stay out of sight and carefully approached from below until my head was almost at a level where I could peek over the pond's rim. Now was the time to use the most extreme caution. Inch by inch, I moved upward until I could see the turtles. So far so good. Then, over the next 20 minutes, I shuffled forward in incremental movements. Eventually, I was positioned in full view of the reptiles, a mere 20 ft (6 m) away, and still had not been detected. Through my telephoto lens, the line of turtles filled the viewfinder and I could plainly see the beautiful red markings on the lower edges of their shells. I was able to get several pictures before one of the turtles noticed the movement of my finger as I pushed the shutter button. In a flash, the log was vacant.

The Lizards

Great Basin Collared Lizard basking on a petroglyph-covered boulder. Pueblo Mountains, southeastern Oregon.

Zebra-tailed Lizard

Callisaurus draconoides

Adult male. Pyramid Lake, northwestern Nevada.

IDENTIFICATION: Adults are 2½–4 in (6–10 cm) in snout-vent length. Males may reach nearly 9 in (23 cm) in total length. This reptile is immediately recognizable by its long, slightly flattened, dark-and-light, crossbanded tail. **The underside of the tail in particular is boldly banded with black and white**. These "zebra-like" markings are prominently displayed when the lizard runs with its tail arched over its body. The legs are very thin and long, with the entire body form streamlined for high speed. The small, smooth, granular dorsal scalation and a countersunk lower jaw facilitate burrowing in sand. The overall upper body coloration is grayish with paired brown blotches (more distinct on females and juveniles), overlaid with bluish-white, light cream, or yellowish spots and mottlings. The sides of the body have dark vertical bars and are tinged with yellow and copper orange. There is a light-edged black stripe at the rear of both hind legs. The throat has a gular fold and a small, slightly extendible, orange dewlap. Males are larger and more robust, and have enlarged postanal scales and prominent, wedge-shaped, black bars and bright blue to greenish-blue patches on the belly. Markings on the belly of females are faint or entirely absent.

VARIATION: Although geographical variation is usually slight and differentiation often difficult, some herpetologists recognize several subspecies, one of which ranges into the Northwest region: the NEVADA ZEBRA-TAILED LIZARD (*C. d. myurus*).

SIMILAR SPECIES: The Long-nosed Leopard Lizard (p. 90), sometimes shares the same habitat and may have similar dark crossbanding on the dorsal surface of its tail. These markings, however, do not extend to the underside of the tail as they do on the Zebra-tailed Lizard.

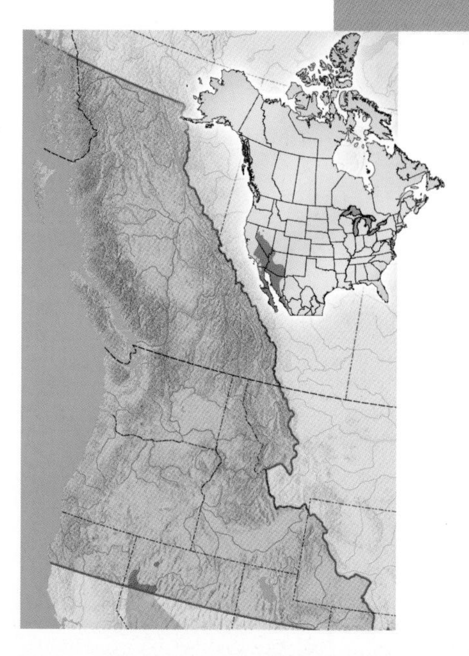

DISTRIBUTION: This colorful lizard reaches the northern limit of its distribution in the desert basins of northwestern Nevada. It is very common around Pyramid Lake at the southern boundary of the area covered by this guide. It has also been recorded nearby in the Winnemucca Lake basin and the vicinity of Lovelock, and it ranges northward into the Smoke Creek Desert. The only other known place the Zebra-tailed Lizard occurs in the Northwest is slightly over 75 mi (120 km) to the northeast near the ghost town of Sulphur in the Black Rock Desert. This is apparently an isolated population, but further field work is needed to determine if this lizard occurs in the intervening area.

HABITAT AND BEHAVIOR: In our region, the Zebra-tailed Lizard is restricted to the warmest arid locations of the northern Great Basin Desert, usually below elevations of 4,500 ft (1,370 m). Typically, it will be seen moving at seemingly rocket-like speed—it

Adult female. Pyramid Lake, northwestern Nevada.

Underside of adult female. Pyramid Lake, northwestern Nevada.

Underside of adult male. Black Rock Desert, northwestern Nevada.

has been clocked at 18 mph (29 km/h)—across open, sandy-gravelly areas and hard-packed washes. For unimpeded running, it requires open places where the shrub and grass cover is widely spaced. Associated plants usually found in its habitat are Big Greasewood, Bailey's Greasewood, Shadscale, Four-winged Saltbush, and Smokebush. If present, stones may be used for basking, even during the hottest times of the day. When threatened, this lizard arches its tail and nervously wriggles it about just prior to running. The striking black and white markings probably attract predators to the tail, which is easily broken off, but will later regenerate. Foods consist of insects and their larvae, spiders, smaller lizards, and the leaves and flowers of plants. Mating takes place in spring (probably in May at our northerly latitude) and 2 to 15 eggs are laid during late June through August.

FIELD NOTES: 26 May 1996, 3:00 p.m. While driving a dirt road near Sulphur in the northeastern arm of the Black Rock Desert, Craig Zuger and I encountered two large male Zebra-tailed Lizards crossing in front of us. I had to jam on the brakes when they nearly shot under our wheels, apparently preoccupied by some sort of territorial dispute. Undoubtedly, it was the mating season for this species in northwestern Nevada. Both individuals displayed especially vivid markings and coloration, and they continued their jousting a short distance off the side of the road. Craig immediately hopped out and began stalking the lizards along the sandy, brushy slope. Using a pole affixed with a fishing-line noose, he soon had one captured. We admired and photographed its beautiful greenish-blue, black, and yellow markings on the belly and sides and its unmistakable "zebra" tail. This encounter was a pleasant surprise, because the previous most northerly records for the Zebra-tailed Lizard was nearly 100 mi (160 km) to the southwest. It was a new range extension and a successful field day.

Juvenile. Pyramid Lake, northwestern Nevada.

Great Basin Collared Lizard

Crotaphytus bicinctores

Adult male in a typical basking pose on a boulder. Pyramid Lake, northwestern Nevada.

OTHER NAMES: Mojave Black-collared Lizard.

IDENTIFICATION: A large lizard, adults are 2½–4½ in (6–11 cm) in snout-vent length. Males sometimes reach a total length of nearly 13 in (33 cm). With its dinosaur-like, **big head, narrow neck, bulky body, large hind legs, and long, thick tail**, this lizard resembles a miniature *Tyrannosaurus rex.* Also distinctive are the **black and white "collar" markings on the neck** and the bright dorsal coloration and pattern. The chocolate brown to olive-brown back has contrasting reddish-orange or yellowish-orange crossbars with a sprinkling of white spots. The feet are yellow. The tail is flattened vertically, and it has brown spots along the sides and an unmarked lighter strip along the top. The scales on the top of the head are only slightly larger than the small granular scales of the body. The belly is primarily white with large dark patches on the lower abdomen and groin. Males are larger, and they have broader heads, a black throat surrounded by blue, and enlarged postanal scales. Females are smaller, lack the blue-black throat, and are usually less brightly colored. During the breeding season, however, gravid females develop intense reddish-orange bars and spots on the body and neck. Juveniles have particularly bright crossbanding.

VARIATION: No subspecies are currently recognized for this lizard, and there appears to be little significant variability within its range in the Northwest.

SIMILAR SPECIES: The Long-nosed Leopard Lizard (p. 90), which sometimes shares the same desert areas, can be as large as the Great Basin Collared Lizard and has a

comparatively broad head, but it lacks the distinctive black and white collar markings on the neck. The Desert Spiny Lizard (p. 94), which is restricted to northwestern Nevada in our region, is also quite large and has similar collar-like markings in the form of a black wedge on each side of the neck. It is easily differentiated, however, by its large, prickly dorsal scales and bright blue patches on the belly.

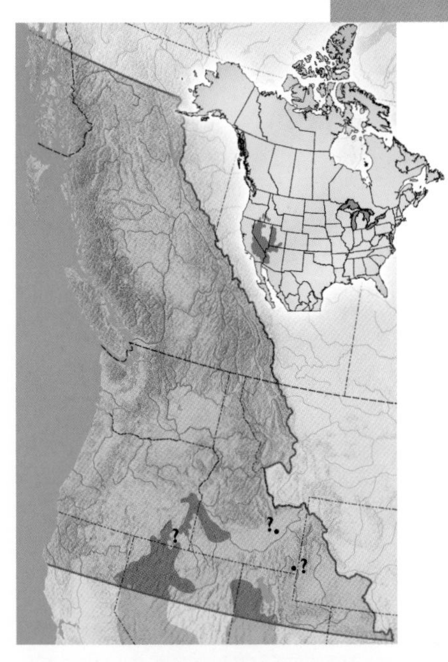

DISTRIBUTION: In the Northwest, the Great Basin Collared Lizard occurs in suitable habitat in the deserts of northwestern Nevada, western Utah, southeastern Oregon, and southwestern Idaho at elevations below 5,000 ft (1,520 m). It has also been recorded in the West Wendover area of northeastern Nevada and in northeastern California's Honey Lake Basin. Two isolated records from southeastern Idaho may constitute relict populations but need verification. One is near Atomic City, Butte County; the other is at Montpelier, Bear Lake County.

HABITAT AND BEHAVIOR: This lizard is found in the arid northern Great Basin Desert ecosystem on loose-soiled slopes that have a sparse growth of low shrubs. Typical associated plants are Shadscale, Big Greasewood, Spiny Hopsage, Bud Sagebrush, and Big Sagebrush. An important component of its habitat is scattered boulders that are used for basking and as lookout stations. This lizard also occasionally uses the tops of small bushes for sunning. It is extremely heat tolerant, and it often basks on rocks during summer's most intense sunshine. When frightened, a Great Basin Collared Lizard will usually dash away and retreat into a hole under a rock or a convenient rodent burrow. It will sometimes run on its hind legs in bipedal dinosaur fashion. During the breeding season, in May and early June, males become aggressively territorial and can often be approached closely. A male will stiffen its legs, bob up and down, sway to and fro, and extend a small blue-and-black dewlap on its throat in an attempt to drive away an intruder. This display is used as both a defense against other invading males and to attract females. If captured, this lizard will usually not hesitate to inflict a

Undersides of male (left) and female (right). Pueblo Mountains, southeastern Oregon.

Great Basin Collared Lizard

Adult female with breeding color. Pueblo Mountains, southeastern Oregon.

Adult male. Pueblo Mountains, southeastern Oregon.

Juvenile. Pyramid Lake, northwestern Nevada.

painful bite with its large, strong jaws. Insects and smaller lizards (including the spine-covered Desert Horned Lizard) are primary foods, along with spiders, small snakes, and the leaves and flowers of various plants. Although our knowledge is meager concerning the reproduction for this species in the Northwest, it is likely that only one clutch of eggs is produced each year at our northern latitude. Records indicate that three to seven eggs are deposited in a burrow, often under a rock. I have observed newly hatched juveniles in late August and early September in southeastern Oregon's Alvord Basin and Owyhee Canyonlands.

FIELD NOTES: 29 May 1991. Doug Calvin and I were relieved to see the day dawn sunny and clear in southeastern Oregon's Alvord Basin. The night had been cold, rainy, and windy. We were acting as guides to a group of Seattle members of the Pacific Northwest Herpetological Society and wanted to show them the typical reptiles of the Great Basin Desert. Although chilly gusts were still blowing, by mid-morning, the sun had warmed things sufficiently for us to find three species of smaller lizards. We were still apprehensive, however, that temperatures would not rise sufficiently to bring out the larger, heat-loving Great Basin Collared Lizard. After lunch, we all drove to a canyon in the Pueblo Mountains that might offer some protection from the wind. Sure enough, after climbing the boulder-strewn slopes, people soon began finding collared lizards. Doug and I zeroed-in on a beautiful male that was in full breeding coloration. Typically territorial, the big lizard held its ground on the top of a rock and posed for photographs. In the past, while kneeling low to take close-up shots, I have occasionally been startled when a fearless male has jumped on my shoulder or the top of my head. After the photo session, Doug noosed the lizard for closer inspection and was bitten as he extracted it from the fishing-line loop. The tenacious reptile held on like a bulldog and, judging from Doug's pained facial expression, illustrated to the other field trip participants that Collared Lizards do have very strong jaws!

Biologist Doug Calvin demonstrating that this lizard can inflict a painful bite. Pueblo Mountains, southeastern Oregon.

Adult male in a defensive pose, extending the dewlap on his throat. Pyramid Lake, northwestern Nevada.

Long-nosed Leopard Lizard

Gambelia wislizenii

Adult male. Owyhee River Canyon, southeastern Oregon.

IDENTIFICATION: Adults are 3½–5½ in (9–14 cm) in snout-vent length. Females may grow to 13 in (33 cm) in total length. This reptile is readily identified by its beautiful **pattern of dark brown "leopard" spots on a light, grayish-tan or golden-tan background. Overlaid on this pattern is a series of pale, creamy-white transverse lines that are often interconnected by smaller mottled lines of the same pale color.** On some individuals, the dark dorsal spots are rather squarish and are so enlarged that the light background is nearly obscured and the pale, creamy-white lines stand out in contrast. Less commonly, the pattern is reduced and simplified to small dark spots on a uniform pale ground color. Long-nosed Leopard Lizards are darker when it is cool and lighter during warmer temperatures. The long, round tail is patterned dorsally with paired dark spots, which often converge to form crossbars. The entire underside of the body is white and **the throat is marked with gray streaks and spots.** Dorsally, the scalation is granular. Males have enlarged postanal scales. Females are larger, and when they are gravid during the breeding season, in late May through June, they develop a bright reddish-orange suffusion on the ventral surface of the tail. Also, orange spots and bars appear on the sides of the body and head. Juveniles have large heads and a checkered dorsal pattern of squared, reddish-brown blotches that are outlined in pale, golden yellow.

VARIATION: This species has an extremely variable dorsal pattern throughout its distribution in the American West. The Long-nosed Leopard Lizard may have large spots in some geographic areas, small spots in other localities, or a mixture of both patterns within a single population. Although not unanimously accepted by all herpetologists, several subspecies have been described. Depending upon the authority consulted, in recent years the Long-nosed Leopard Lizards in our region have been assigned to two different subspecies: the LARGE-SPOTTED

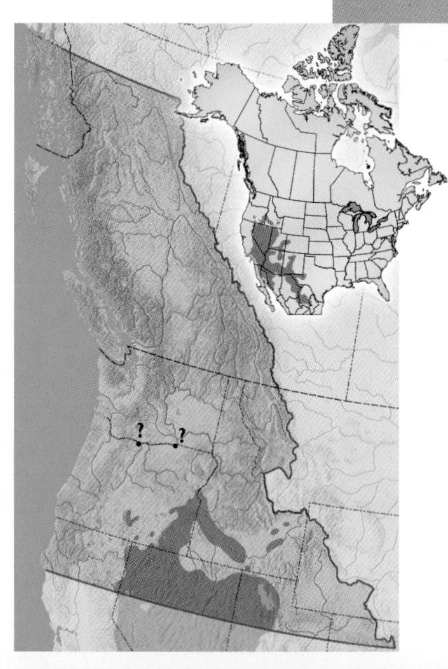

LEOPARD LIZARD (*G. w. wislizenii*) or the LAHONTAN BASIN LEOPARD LIZARD (*G. w. maculosus*).

SIMILAR SPECIES: The closely related Great Basin Collared Lizard (p. 86) often shares the same desert areas and is superficially similar in size and shape. However, the distinctive black and white collar markings on the neck will quickly differentiate it from the Long-nosed Leopard Lizard.

DISTRIBUTION: The Long-nosed Leopard Lizard ranges northward from the American Southwest through much of Nevada, western Utah, northeastern California (Honey Lake Basin and Surprise Valley), southeastern Oregon, and southern Idaho at elevations below 5,000 ft (1,520 m). Isolated populations occur in the vicinities of several dry lake beds in Oregon's Lake County, and in the eastern portions of Idaho's Snake River Valley. There are also old records for Oregon

Adult female with breeding color. Alvord Basin, southeastern Oregon.

Long-nosed Leopard Lizard

Underside of a female, showing dark gray markings on the throat and orange breeding color on the tail. Alvord Basin, southeastern Oregon.

Juvenile. Pyramid Lake, northwestern Nevada.

along the Columbia River at The Dalles, Wasco County, and for Hat Rock, Umatilla County. The Umatilla record is represented by a specimen taken in 1959 and preserved in the Oregon State University collection, but the Wasco record is thought to be from the late 1800s or the early 1900s, and it may be based on erroneous data. Field work is needed to verify if this species still inhabits the Columbia Basin.

HABITAT AND BEHAVIOR: This lizard occupies sandy-gravelly desert flats, brushy dune systems, and loose-soiled, open hillsides in the Northwest. An aggressive predator, it favors places with open expanses between bushes that allow space for running and foraging. During the hot part of the day, it often hides under a shady shrub, waiting to ambush unsuspecting prey. Where stones are present in the habitat, they are often used for basking, as are the tops of bushes. Look for this lizard among the sand hummock mounds that collect around Big Greasewood. A startled Long-nosed Leopard Lizard will dash under a bush for protection and remain motionless, its spotted pattern blending with the dappled shadows. If pursued further, it will usually seek refuge in a rodent burrow. Fast and wary, the Long-nosed Leopard Lizard is capable of running on its hind legs. If captured, it will hiss (occasionally producing a squealing sound), bite, and open its mouth to display a black throat lining. This voracious feeder will often eat animals almost as large as itself. Other lizards are devoured (sometimes including its own species), along with insects, spiders, and the leaves and flowers of various plants. Small snakes and pocket mice have also been recorded as food items. In June or July, females deposit a clutch of three to seven eggs in a self-excavated burrow. Hatching takes place during August or early September.

FIELD NOTES: 31 July 1997. On a perfect, blue-skied, high desert summer morning,

Adult basking in late evening sunshine on a Shadscale bush. Alvord Basin, southeastern Oregon.

long-time friend and wildlife artist, Craig Zuger, and I decided to explore the sandy, brushy slopes above our camp by Pyramid Lake in northwestern Nevada. We hoped to find lizards out doing their morning basking and foraging before the sun became too intense for them to be as active. We slowly worked our way upward and across the slope, our senses alert for lizards moving between bushes or sunning on rocks. At about 9:00 a.m., Craig shouted, "Al, here's a lizard eating another lizard!" Electrified, I quickly clambered up the hill, arriving to see a large, female Long-nosed Leopard Lizard, directly in front of a grinning Craig. It had just captured a small female Zebra-tailed Lizard, held crosswise in its strong jaws. I slowly knelt to the ground and gingerly positioned my camera and tripod at a low angle, trying to avoid disrupting this reptilian breakfast. Over the next 20 minutes I focused my telephoto macro lens on the Long-nosed Leopard Lizard as it methodically worked the Zebra-tailed Lizard around in its mouth until its victim could be swallowed head-first. Subsequently, I got an entire series of shots as the prey species disappeared down the gullet of the predator species. What a rare and fortunate nature photography opportunity.

Adult, showing how its spotted dorsal pattern blends with dappled shade under a Big Greasewood bush. Summer Lake, southeastern Oregon.

Adult eating a Zebra-tailed Lizard. Pyramid Lake, northwestern Nevada.

Desert Spiny Lizard

Sceloporus magister

Adult male in a typical basking pose. Pyramid Lake, northwestern Nevada.

IDENTIFICATION: Adults are large, 3½–5½ in (9–14 cm) in snout-vent length and to nearly 13 in (33 cm) in total length. **The body form is stout, with extremely large, pointed dorsal scales that have a noticeable sheen.** The general impression given is of **a big, robust, prickly lizard that has a distinctive black, wedge-shaped mark on each side of the neck.** Dorsally, the overall coloration is grayish brown, with a scattering of yellow scales (especially on the sides). The back usually has a pattern of dark brown crossbars and blotches, with a row of black spots along the sides. Younger specimens have well-defined dorsal crossbars, but older adults may sometimes tend to be an overall yellowish brown with obscured markings. **Males have large, vivid, turquoise blue patches on the throat and both sides of the belly**, and enlarged postanal scales. On females, the blue patches are faint or entirely absent, and a reddish-orange coloration on the head develops during the breeding season (May and early June).

VARIATION: Several subspecies have been described, with one occurring in our area: the YELLOW-BACKED SPINY LIZARD (*S. m. uniformis*).

SIMILAR SPECIES: The closely related Western Fence Lizard (p. 98), which often shares the same habitat, has smaller, less prickly dorsal scales, lacks the black wedge shapes on the neck, and reaches a maximum snout-vent length of only about 3½–4 in (9–10 cm). The Great Basin Collared Lizard (p. 86) has similar black markings on the neck, but the dorsal scales are granular rather than spiny.

DISTRIBUTION: The Desert Spiny Lizard enters the region covered by this guide only in

Adult female. Pyramid Lake, northwestern Nevada.

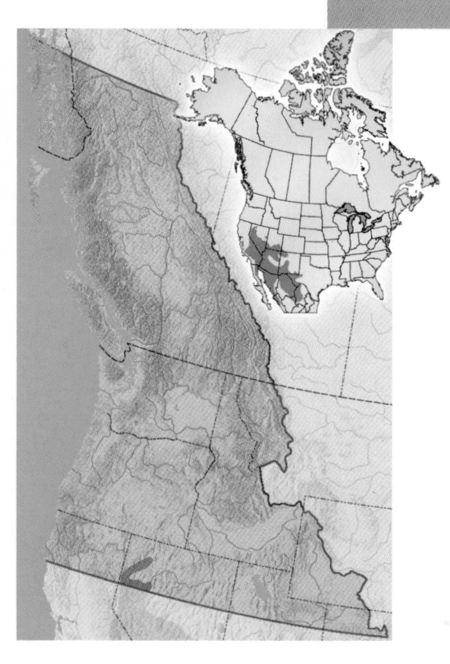

northwestern Nevada at elevations below 4,500 ft (1,370 m). It is common around Pyramid Lake and is also found in the Winnemucca Lake basin and the Lovelock area. Additionally, this species ranges northward into the Black Rock Desert, at least to the vicinity of the ghost town of Sulphur.

HABITAT AND BEHAVIOR: These sun-loving lizards inhabit the exposed drylands of the northern Great Basin Desert. Although they are occasionally found on the ground or climbing shrubs, trees, and wooden fence posts, the favored habitat seems to be rocky slopes and outcrops. Large numbers can be seen basking on the tufa formations along the edges of Pyramid Lake. Big and conspicuous, they are easily spotted from a moving car as one drives through rocky

Older adult with obscured markings. Pyramid Lake, northwestern Nevada.

Adult female. Pyramid Lake, northwestern Nevada.

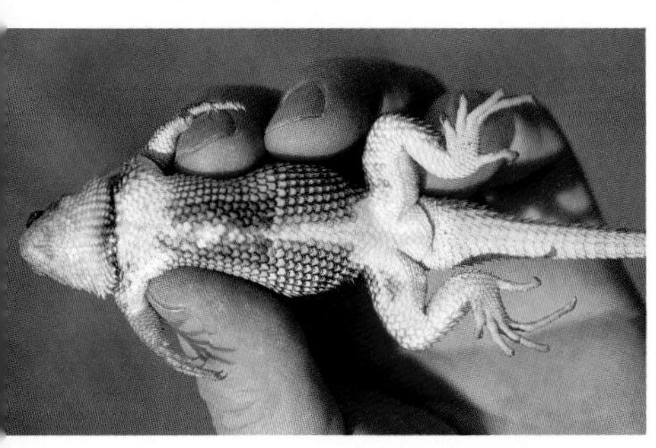

Underside of an adult male. Pyramid Lake, northwestern Nevada.

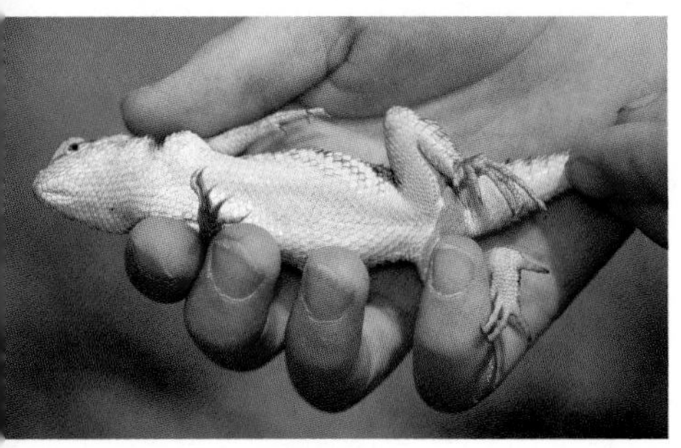

Underside of an adult female. Pyramid Lake, northwestern Nevada.

places. If approached, a Desert Spiny Lizard will usually try to hide by warily shifting its position to the opposite side of a rock and finally darting into a deep, safe crevice. If the observer is patient and remains motionless, the reptile will often reappear in a few minutes. The diet includes insects and their larvae, spiders, centipedes, smaller lizards, and occasionally the leaves and flowers of plants. Very little is known about the breeding habits of this lizard in the Northwest. Like many desert lizard species at our northern latitude, probably only a single clutch of eggs is deposited each year in mid-July or early August, with hatching taking place in late August or early September. Records from the southern portions of its range indicate 4 to 19 eggs are laid in each clutch.

FIELD NOTES: 5 September 1998. After arriving on the previous windy and rainy night at Pyramid Lake in northwestern Nevada, the morning broke clear and sunny. My wife, Jan, our young son, Matthew, and friends Laura McMasters and her grandson, Sebastian, were all eager to help look for reptiles. The sunshine rapidly warmed the rocks by 9:00 a.m., and we optimistically set out across the slopes above the brilliantly blue-green lake. Both of the boys were equipped with lizard nooses, because I needed photos of the belly coloration of the female Desert Spiny Lizard. Kids love noosing lizards as much as fishing, and I had faith that they would capture specimens in short order. I was not disappointed. Within a few minutes, Sebastian exclaimed that he had spotted a big Desert Spiny Lizard sunning on a rock and proceeded to stalk it. Only moments later, Matthew gave an excited shout that he'd located the same species. Typical of the Desert Spiny Lizard, the reptiles scooted around to the opposite sides of the rocks as the boys approached. Matthew and Sebastian both slowly circled around their respective rocks, making earnest attempts to snare the wary creatures. In no time at all, they proudly presented me with captured lizards. As it turned out, both specimens were females, I got my photographs, and the lizards were released.

Juvenile. Pyramid Lake, northwestern Nevada.

Large numbers of Desert Spiny Lizards can be observed basking on the tufa formations around Pyramid Lake in northwestern Nevada.

Western Fence Lizard

Sceloporus occidentalis

Adult male Great Basin Fence Lizard. Pueblo Mountains, southeastern Oregon.

IDENTIFICATION: Adults are usually 2½–3½ in (6–9 cm) in snout-vent length and to nearly 7 in (18 cm) in total length. Individuals in populations east of the Cascades may sometimes slightly exceed the average size. This reptile is well known as the "blue-bellied lizard" because of the **large, solid blue patch on each side of the abdomen, with a matching patch of blue on the throat.** The dorsal coloration is primarily grayish brown, with a checkered pattern of darker, triangular blotches or crossbars on the back, often interspersed with white flecks. There is a yellow or orange tinge on the undersides of the front and hind legs. **The dorsal surface of the body has a prickly look and feel because of its rather large, pointed scales**. Males have larger, more intense blue patches on the throat and belly, with black outlining the blue areas. There is usually a scattering of turquoise blue flecks on the back as well, and the postanal scales are enlarged. The blue patches of the belly and throat on females are faint or occasionally entirely lacking, and there are no turquoise blue flecks in the dorsal pattern. Very dark, gray-brown to entirely black specimens of both sexes are common in the Great Basin region (occasionally with dorsal mottlings of bright yellow or rusty orange on the body, tail, and legs, and around the eyes).

VARIATION: Six subspecies have been described for this lizard, with two occurring in our region. The NORTHWESTERN FENCE LIZARD (*S. o. occidentalis*) is smaller, generally less than 3½ in (9 cm) in snout-vent length. Its mid-belly is usually white or pale gray, with scattered dark flecks and narrow black outline around blue patches. The dorsal coloration is generally lighter gray-brown, with a checkered pattern of darker triangular blotches. This subspecies occurs primarily west of the Cascades, but it ranges along the eastern slopes of these mountains in central Washington and into the lower Deschutes River drainage of north-central Oregon. The GREAT BASIN FENCE LIZARD (*S. o. longipes*) attains a larger size of nearly 4 in (10 cm) in snout-vent length. Its mid-belly is gray, with an extensive black suffusion around the blue patches. The dorsal coloration is darker, with

a pattern of narrow, irregular crossbars. Completely black individuals are common. This subspecies is found east of the Cascades as far north as southeastern Washington.

SIMILAR SPECIES: The closely related Sagebrush Lizard (p. 106) coexists with the Western Fence Lizard east of the Cascades and in the Klamath Mountain region of southwestern Oregon and northern California. It is smaller, however, rarely exceeding 2½ in (6 cm) in snout-vent length, and the dorsal pattern often tends to be arranged in lengthwise stripes (especially east of the Cascade Mountains). The Sagebrush Lizard also has smaller, less prickly dorsal scales, and smooth granular scales on the rear of the thighs, and males have mottled blue on the throat instead of a solid blue patch. The Desert Spiny Lizard (p. 94) of northwestern Nevada has larger, pointed dorsal scales and black, wedge-shaped marks on the neck, and it is more robust and larger, being up to 5½ in (14 cm) in snout-vent length. The Eastern Fence Lizard (p. 102) is very similar in appearance, but the ranges of the two species do not overlap in our region.

DISTRIBUTION: West of the Cascades, this species is abundant in the valleys, mountains, and coastal areas of northwestern

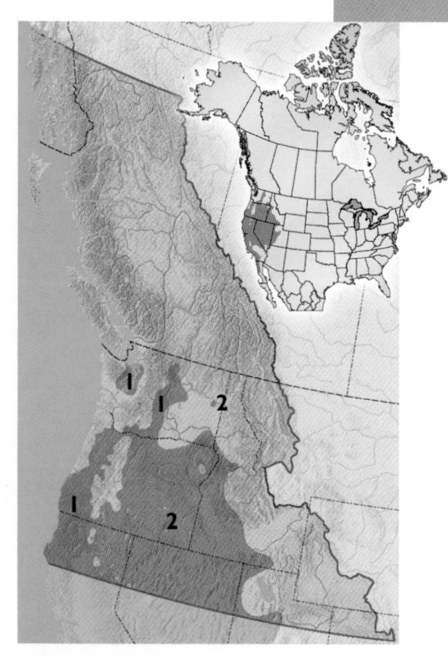

SUBSPECIES

- 1 Northwestern Fence Lizard *S. o. occidentalis*
- 2 Great Basin Fence Lizard *S. o. longipes*
- Species range outside the Pacific Northwest

Black variation of the Great Basin Fence Lizard. Black Rock Desert, northwestern Nevada.

Western Fence Lizard

Adult female Northwestern Fence Lizard. The Nature Conservancy's Dye Creek Preserve, near Red Bluff, Sacramento Valley, northern California.

Underside of a male Great Basin Fence Lizard, showing the gray belly typical of this subspecies. Alvord Basin, southeastern Oregon.

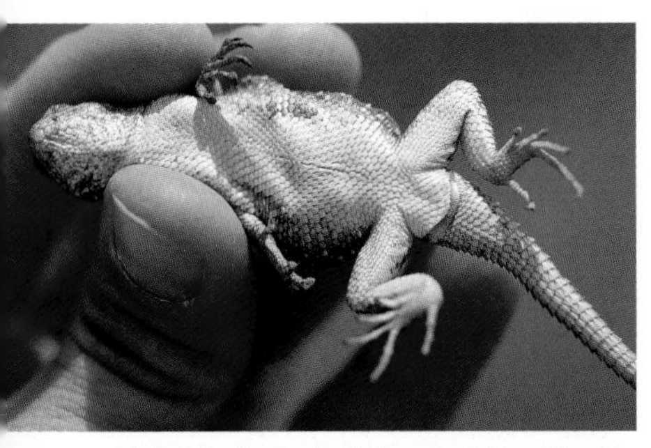

Underside of a female Northwestern Fence Lizard, showing the white belly typical of this subspecies. The Nature Conservancy's Dye Creek Preserve, near Red Bluff, Sacramento Valley, northern California.

California and southwestern Oregon. It ranges northward through Oregon's interior Umpqua River drainage and the Willamette Valley. This lizard also occurs in the Columbia River Gorge and as scattered populations in northwestern Washington near the Puget Sound–Hood Canal shorelines. East of the Cascade Mountains, it inhabits most elevations below about 5,000–6,000 ft (1,520–1,830 m) in northeastern California, northern Nevada, northwestern Utah, southwestern Idaho, central and eastern Oregon, and central and southeastern Washington. The Western Fence Lizard penetrates fairly high up both sides of the Cascade Mountains on the open, rocky, southern slopes of some canyons.

HABITAT AND BEHAVIOR: This reptile is probably the most commonly seen lizard throughout much of the Northwest—it is absent only from shady, dense forests and extremely arid desert flats. It can be observed basking on wooden fence posts, stumps, logs, and rocks on sunny, south-facing hillsides in the oak woodlands west of the Cascade Mountains. In the dry interior plateau country, the Western Fence Lizard is a resident of canyons, rimrocks, and boulder covered slopes in deserts and open juniper and pine forests. An excellent climber, it will often seek

refuge by ascending a tree and nimbly avoiding the pursuer by going around and around the trunk, always attempting to stay on the opposite side. The diet consists mainly of insects and spiders. Depending upon the elevation, mating takes place in late April to mid-May, with 3 to 17 eggs laid between late May and early July. The hatchlings appear in August and early September.

Adult Great Basin Fence Lizard. Alvord Basin, southeastern Oregon.

FIELD NOTES: 14 April 1990. Doug Calvin and I spent a warm afternoon hiking in a canyon near Rowena on Oregon's side of the Columbia River Gorge. It felt good to be out after the winter, searching through the boulders on this oak-covered slope. With a yawn, however, Doug announced that he had gotten too little sleep the night before. After stretching out on a sun-warmed log, he was soon napping. Continuing to hunt for reptiles about 20 ft (7 m) away, I noticed two male Western Fence Lizards that were sparring on some rocks. It quickly became apparent that a nearby female was triggering this macho confrontation. The two Romeos attempted to intimidate each other with displays of bobbing up and down and tilting the sides of their bodies upward to display their intensely blue bellies. Both males began to wildly pursue each other among the rocks, with the chase at one point spiraling up and down a tree trunk. They were so engrossed in their brawl that I was able to move in closely with my camera and get many good pictures. Eventually, they ended up skittering about on the same log with Doug's inert body. Once or twice, the reptiles ran across his legs or nearly crawled onto his head. My dozing friend remained oblivious, while the female lizard seemed totally disinterested in her passionate suitors. Soon, the conquering male, his jaws grimly locked on the neck of his adversary, pinned the other lizard's back to the ground, the loser's blue belly turned to the sky. The vanquished contestant wriggled free and escaped. Doug awoke, frightening away all three lizards as he sat up and naively asked, "Seen any reptiles?"

Male Western Fence Lizards fighting. Columbia River Gorge National Scenic Area, near Rowena, Oregon.

Eastern Fence Lizard

Sceloporus undulatus

Adult male; grayish-tan variation. Near Duchesne, Uinta Basin, northeastern Utah.

IDENTIFICATION: Adults are 2–3½ in (5–9 cm) in snout-vent length and to 7½ in (19 cm) in total length. The dorsal coloration and pattern is somewhat variable, but it is commonly grayish tan to reddish tan. Some individuals are reddish brown on the back, with contrasting, lighter, grayish tan on the sides of the body. Others have a pattern of narrow, wavy, dark lines that cross the back. **Both sexes have a relatively narrow, elongated patch of blue on each side of the belly, with a wide white area between them. There is a small, solid patch of blue at both rear corners of the throat, widely separated by white.** The dorsal scalation consists of rather large, pointed scales, giving the lizard a prickly appearance. Males have enlarged postanal scales, and usually a narrow black edging (sometimes faint) along the inner side of the blue belly patches.

VARIATION: Nine subspecies of this lizard are currently recognized. Only one is found in our area: the Northern Plateau Lizard (*S. u. elongatus*).

SIMILAR SPECIES: The Eastern Fence Lizard often shares its habitat with the following two species of lizards that have similar dorsal markings and blue patches on their bellies and throats. The Sagebrush Lizard (p. 106) is a close relative, but adults are smaller, reaching snout-vent lengths of only 2–2½ in (5–6 cm). It also differs in having smoother, less prickly dorsal scales and blue mottling on the throat instead of two small

solid blue patches. The Ornate Tree Lizard (p. 114) is easily identified by its gular fold on the throat and a body that is primarily covered with small, granular scales, unlike the Eastern Fence Lizard's large, prickly dorsal scales.

DISTRIBUTION: This reptile has a limited range within the area covered by this guide. It is found only in the Green River drainage of northeastern Utah and northwestern Colorado and in a small section of southwestern Wyoming in the vicinity of Flaming Gorge. At our northern latitude, it probably does not occur above elevations of 6,500 ft (1,980 m).

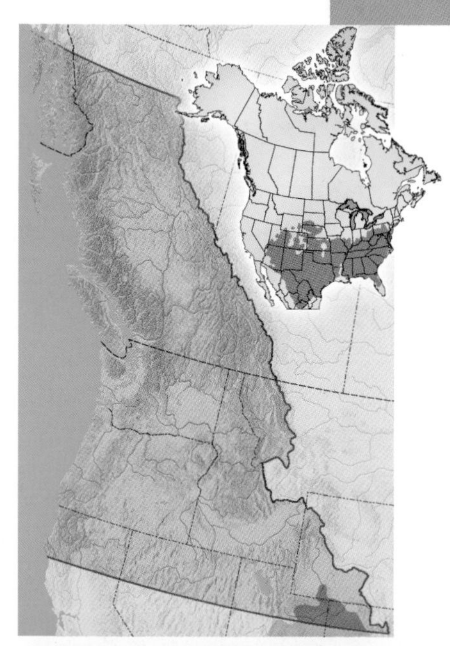

HABITAT AND BEHAVIOR: The Eastern Fence lizard occurs in a variety of habitats across its extensive distribution, which reaches to the Atlantic Coast. Within the relatively small portion of its range that includes the Northwest, it is primarily an inhabitant of rocky canyons, escarpments, and boulder-strewn slopes among juniper woodlands. In some places, it occurs in shrubby basins and open grasslands, but rock outcrops are always present. Food items recorded for this species include adult and larval insects, as well as spiders, ticks, and millipedes. There is also one reported observation of a juvenile Sagebrush Lizard having been eaten. Little is known of the breeding habits of the Eastern Fence Lizard in our area. Field studies on the eastern slopes of the Rockies in central Colorado have found that mating takes place in late May and early June. Females there lay two yearly clutches of an average eight eggs during late June or early July. Hatching takes place between mid-July to early September.

Adult male; variation with a reddish-brown back. Near Duchesne, Uinta Basin, northeastern Utah.

FIELD NOTES: 12 September 1997. I spent the morning hiking along a south-facing rocky rim near Duchesne in the Uinta Basin of northeastern Utah. At about 10:00 a.m. I began to see lizards sunning on the tops of boulders, ledges, dead tree limbs, and logs. Initially, "Western Fence Lizard" flashed through my mind, because that species was the familiar one first imprinted on my youthful memory in Oregon's Willamette Valley. I immediately realized my mistake, though, because that species does not range east of the Wasatch

Adult female; reddish-tan variation. Near Vernal, Uinta Basin, northeastern Utah.

Gravid female. Near Duchesne, Uinta Basin, northeastern Utah.

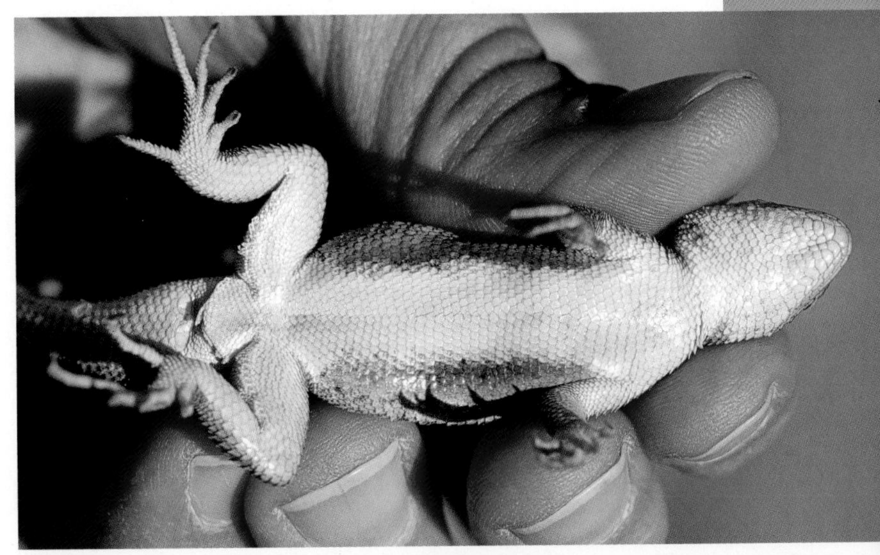

Underside of a male. Near Duchesne, Uinta Basin, northeastern Utah.

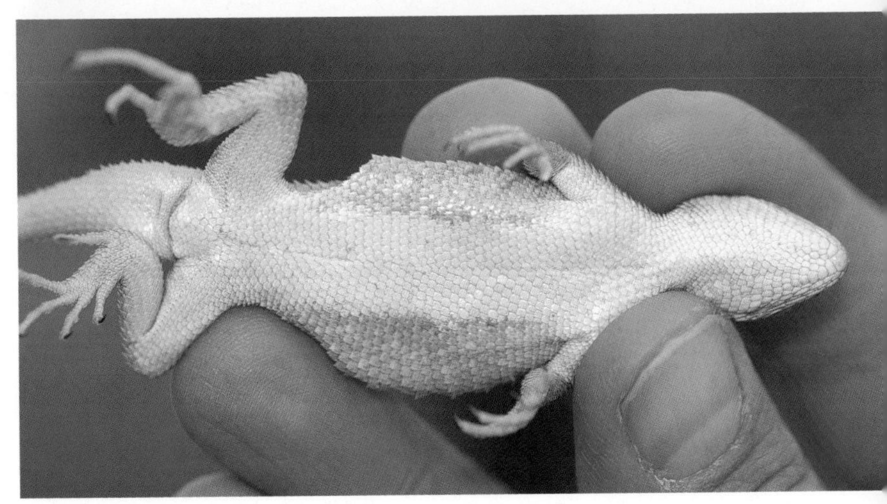

Underside of a female. Near Vernal, Uinta Basin, northeastern Utah.

Mountains in northern Utah. These reptiles were undoubtedly Eastern Fence Lizards. With my lizard noose I soon captured a specimen and made a quick confirmation. The small patch of blue at both rear corners of the throat were quite different from their western relative's solid blue throat patch. However, the similarities in general appearance, body movements, and choice of basking stations were uncanny. After releasing the lizard, I sat on a rock and contemplated the very different overall habitat. Most notably, there are the reddish sandstone formations so typical of the Colorado River canyonlands of southern Utah. Surrounding these scenic rocks is a sparse forest of short, twisted Utah Juniper, along with several species of small cacti and yucca growing on the reddish, sandy soil. This area looks like the American Southwest annexed north.

Sagebrush Lizard

Sceloporus graciosus

Adult male Northern Sagebrush Lizard. Black Rock Desert, northwestern Nevada.

IDENTIFICATION: A small lizard, adults are 2–2½ in (5–6 cm) in snout-vent length and to slightly over 5½ in (14 cm) in total length. The dorsal coloration of the mid-back is brown or grayish brown, bordered on each side by two darker, lengthwise stripes (or rows of partially connected dark blotches), which, in turn, are divided by a pale gray or tan stripe. The sides of the body and neck often have a rusty-orange hue. **Unlike in other members of the *Sceloporus* genus, the dorsal scales are not as large, pointed, or prickly. The Sagebrush Lizard's small scales impart only a slightly rough texture to the back.** Males have an elongated patch of blue on each side of the abdomen, **a mottled blue pattern on the throat**, and enlarged postanal scales. Females usually lack blue markings on the belly and throat or have only very faint bluish-gray markings. During the breeding season in April and May, there is an intensification of the rusty-orange coloration on the neck and sides of females.

VARIATION: Three subspecies are currently recognized, with two ranging into the Northwest: the NORTHERN SAGEBRUSH LIZARD (*S. g. graciosus*), which occurs east of the Cascade Mountains, has a dorsal pattern of distinct lengthwise stripes; the WESTERN SAGEBRUSH LIZARD (*S. g. gracilis*), which occurs west of the Cascade Mountains, has some or all of the dark dorsal stripes reduced to rows of blotches or partially connected blotches.

SIMILAR SPECIES: Juveniles or small individuals of the closely related Western Fence Lizard (p. 98) and Eastern Fence Lizard (p. 102) could easily be mistaken for this species. A male Sagebrush Lizard, however, has a mottled blue pattern on the throat,

whereas the Western Fence Lizard has a solid blue throat patch (sometimes divided into two sections), and the Eastern Fence Lizard has a small, solid blue patch at both rear corners of the throat. Fence lizards' large, pointed, prickly scales on the body and rear of the thighs are also different from the Sagebrush Lizard's slightly rough-textured dorsal scalation and the smooth, granular scales on the rear of the thighs. The Common Side-blotched Lizard (p. 110) is similar in size and appearance and often shares the same habitat east of the Cascades. It can be distinguished by its smooth, granular dorsal scales, a gular fold at the rear of the throat, and an absence of blue markings on the belly in males.

DISTRIBUTION: The Sagebrush Lizard is found in suitable habitat throughout most of the drylands east of the Cascade Mountains in northern Nevada, northeastern California, central and eastern Oregon, central and southeastern Washington, southern Idaho, western Wyoming, northwestern Colorado, and northern Utah. It also occurs west of the Cascades in the Klamath Mountain area of northwestern California and southwestern Oregon. This species often ranges to high elevations—it has been recorded at 7,000 ft (2,130 m) in Crater Lake National Park, Oregon, and to 8,300 ft (2,530 m) in geothermally warmed areas in Yellowstone and Grand Teton national parks in western Wyoming.

HABITAT AND BEHAVIOR: As its common name suggests, this lizard is found in sagebrush ecosystems over a large portion of its distribution. Less commonly, it is seen in the arid saltbush-greasewood communities of the Great Basin Desert. At higher elevations it will be encountered in dry, open juniper and pine forests where the predominant shrub cover is Bitterbrush. In the Klamath Mountains, this reptile is an inhabitant of chaparral areas, often where thinly vegetated serpentine soils support a plant association of Jeffrey Pine, manzanita species, Western Azalea, and Buckbrush. It is the most common lizard in this type of habitat in the Kalmiopsis Wilderness of southwestern Oregon. Regardless of the ecosystem, the

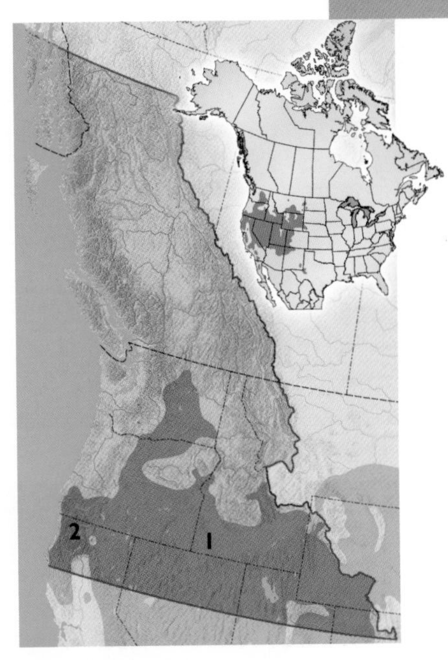

SUBSPECIES

1	Northern Sagebrush Lizard *S. g. graciosus*
2	Western Sagebrush Lizard *S. g. gracilis*
	Species range outside the Pacific Northwest

Adult female Western Sagebrush Lizard. Kalmiopsis Wilderness, Siskiyou Mountains, southwestern Oregon.

Underside of an adult male. Fort Rock State Park, central Oregon.

Underside of an adult female. Shasta Valley, northern California.

Sagebrush Lizard prefers places with open, sunny areas between bushes, usually with loose, sandy soil. If rocks and dead logs are present in the habitat, they are usually used for basking. It has also been observed climbing into the lower branches of shrubs. Favored foods are insects, spiders, ticks, mites, and small scorpions. A clutch of two to seven eggs is laid in June or July and the hatchlings usually appear in August and September. There is a record of hatching taking place as late as October, however, at an elevation of 6,000 ft (1,830 m) in Lassen Volcanic National Park in northern California.

FIELD NOTES: 28 May 1995. I took a quick outing today to Fort Rock State Park in the High Desert country of Oregon's Lake County. The huge, amphitheater-like interior bowl of this natural rock fortress is a good place to photograph Sagebrush Lizards. Even on cool spring days they are usually out because the walls of the "fort" offer considerable wind protection, making for warmer temperatures. However, this particular afternoon presented the opposite problem. Temperatures had unexpectedly soared and it was too hot. Although I saw many lizards, they were all speeded-up by solar energy. With the little reptiles darting from bush to bush across the sandy interspaces, no photo opportunities were presented. Whenever I focused my telephoto macro lens on one that had stopped moving, it was invariably located

deep under a shadowy bush with too many branches in the way. The Sagebrush Lizards were adeptly fine-tuning their body temperatures by constantly flitting from cool shade to hot sunny spots. Through this method and other behavior and adaptations, ectothermic lizards are often able to maintain body temperatures that are only a bit more variable than that of endothermic birds and mammals. Indeed, it is now known that when lizards have bacterial infections they will engage in extended basking to produce a healing "behavioral fever." Learning about these reptilian traits of thermoregulation and planning accordingly in advance can often make the difference between success or failure for a herpetologist. If I'd known the day was going to be so hot, I would have gotten there during the cooler morning hours. The life of a reptile is directly tethered to the sun and, consequently, so is the life of anyone attempting to study them.

Adult basking on a sandstone formation. Red Fleet State Park, Uinta Basin, northeastern Utah.

Common Side-blotched Lizard

Uta stansburiana

Adult male. Black Rock Desert, northwestern Nevada.

IDENTIFICATION: Adults are 1½–2½ in (4–6 cm) in snout-vent length and to 5 in (12 cm) in total length. These small, grayish-brown lizards have thin, rather flat bodies and a **distinctive black blotch on each side of the body behind the front legs (sometimes faint or absent)**. There is usually a speckled dorsal pattern of light and dark spots along the back. The belly is white or pale bluish gray. Often, there is a bluish mottling on the throat, which is **bordered at the rear by a gular fold. The scalation of the entire body is granular**. Males are larger, have slightly enlarged postanal scales, and are more colorful, with turquoise blue spots on the back and a bright reddish-orange suffusion on the chin, sides of the body, and tail. Females have only whitish or very pale blue dorsal spots,

little, if any, orange coloration, and the dark side blotches are smaller and less defined.

VARIATION: Several subspecies have been described, but they are currently not accepted by many experts. Further studies that provide clarification for this species are needed. There are considerable differences in coloration and pattern throughout the overall range of the Common Side-blotched Lizard, even within a single population. Many individuals from the Green River drainage of northeastern Utah and northwestern Colorado have no dorsal spots and markings and lack the orange coloration on the sides of the body (or if this pattern is present, it is very faint.)

SIMILAR SPECIES: The Sagebrush Lizard (p. 106) and juveniles of the Western Fence Lizard (p. 98) and Eastern Fence Lizard (p. 102) could be mistaken for this species. They can be easily differentiated from the Common Side-blotched Lizard by their rough or prickly dorsal scales, lack of a gular fold on the throat, and patches of blue on the bellies in the males. The similar Ornate Tree Lizard (p. 114), which is found in our area only in northeastern Utah, northwestern Colorado, and southwestern Wyoming, has a narrow strip of enlarged scales running down the middle of the back, unlike the Common Side-blotched Lizard's uniformly granular dorsal scales. Also, the Ornate Tree Lizard lacks the black blotch on each side of the body behind the front legs, and males have blue patches on the belly.

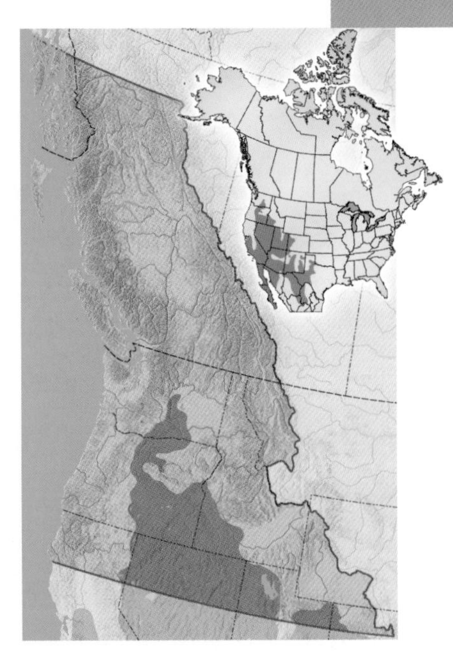

DISTRIBUTION: Occurring throughout a good portion of the interior plateau country of the Northwest, the Common Side-blotched Lizard is found in northern Nevada, northeastern California, northern Utah, northwestern Colorado, southwestern Idaho, southeastern and central Oregon, and south-central Washington.

HABITAT AND BEHAVIOR: The Common Side-blotched Lizard inhabits the arid to

Adult female. Pyramid Lake, northwestern Nevada.

Adult male; pale Colorado Plateau variation. Near Duchesne, Uinta Basin, northeastern Utah.

Adult male, showing the dark blotch on the side of the body. Fort Rock State Park, central Oregon.

semi-arid lands of our region, occurring in both open deserts and sparse juniper woodlands at elevations below 6,000 ft (1,830 m). It is equally at home skittering between bushes on sandy flats as it is perching on boulders jutting from a hillside or rocky canyon. This species is usually the first lizard out and moving in the morning, because its thin, flat body warms quickly. As the temperature rises, it becomes increasingly active, jumping from rock to rock with great agility. A male will often sit atop a boulder and bob up and down, appearing to be doing push-up exercises. In reality, this is a territorial display meant to ward off intruders. This lizard can often be approached closely, seeming to be almost curious about one's actions. The diet includes insects, spiders, ticks, mites, and small scorpions. The spring mating season occurs from April into June. At our northerly latitude, one or two clutches of one to five eggs are deposited each year during May through June. These hatch in late July and early August.

FIELD NOTES: 29 July 1995. The view of towering Steens Mountain was awesome from the level expanse of the Alvord Desert in southeastern Oregon. Craig Zuger and I had camped near Alvord Hot Spring, and after a steamy soak at dawn, we were now driving across the barren, shimmering alkali to reach Big Sand Gap. An Oregon Public Broadcasting crew was meeting us the next day to do a television show about desert reptiles, so we were scouting for photogenic lizards and

snakes. After reaching the eastern edge of the playa, it was only a ½-mile (800-m) hike across brushy dunes to reach the sand-swept mouth of the canyon. We separated, taking routes up opposite sides of the rocky aperture. Within minutes, I spotted a male Common Side-blotched Lizard basking on a boulder. His blue-speckled back and salmon-colored sides glowed in the clear morning sunshine. A short distance further up the slope, a female of the same species zipped under a bush. Within 20 minutes I had reached the upper end of Big Sand Gap and had counted 13 of these diminutive lizards. Pausing to catch my breath, I remembered reading about a field study showing that the home ranges of Common Side-blotched Lizards are smaller than that of most other species. They often confine their movements to a radius of less than 50 ft (15 m), which explains why so many can be encountered on a relatively short hike. Craig called to me from across the canyon that he'd not only found lots of *Uta*, but a nice Great Basin Collared Lizard as well. This location should be great to shoot the segment for the *Oregon Field Guide* show.

Underside of an adult male. Fort Rock State Park, central Oregon.

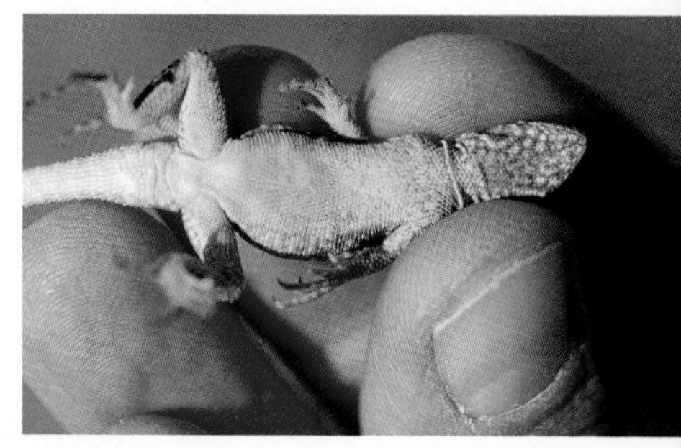

Underside of an adult female. Fort Rock State Park, central Oregon.

Common Side-blotched Lizards are usually the first reptiles out sunning in the morning. Adult male. Alvord Basin, southeastern Oregon.

Ornate Tree Lizard

Urosaurus ornatus

Adult in a characteristic basking pose. Red Fleet State Park, Uinta Basin, northeastern Utah.

IDENTIFICATION: Adults are 1½–2¼ in (4–6 cm) in snout-vent length and to nearly 6 in (15 cm) in total length. This small lizard's dorsal coloration generally matches the surrounding environment, ranging from gray, tan, brown, or reddish brown, to nearly black. The back is usually marked with a series of darker, narrow crossbars, often with a scattering of small, light spots along the sides. The tail is patterned with similar dark crossbars. There is a fold of skin along each side of the body and a gular fold at the rear of the throat. **The dorsal scalation is granular, except for a narrow strip of enlarged scales running down the middle of the back.** Males have bright blue or greenish-blue patches on each side of the belly, a blue throat (sometimes greenish-yellow), and

enlarged postanal scales. Females lack the blue patches on the belly and have a white or orangish-yellow throat.

VARIATION: There is considerable variability in coloration and pattern, both geographically and within a given population. Several subspecies have been described, but they are not accepted as valid by many herpetologists. The CLIFF TREE LIZARD (*U. o. wrighti*) is the race considered to occur in our region.

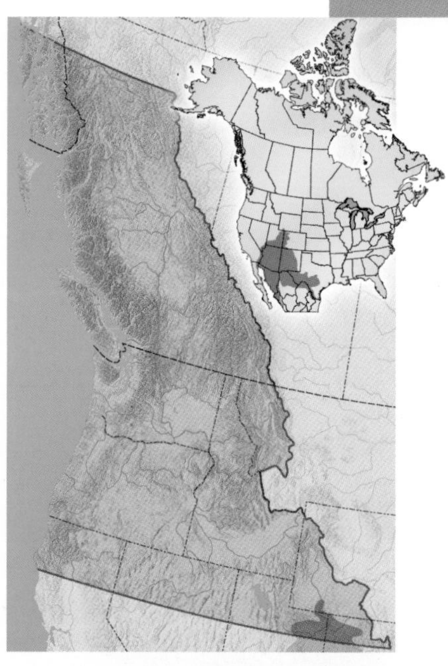

SIMILAR SPECIES: The Common Side-blotched Lizard (p. 110) often shares the same habitat but differs in lacking the strip of enlarged scales down the middle of the back; instead, the dorsal scales are entirely granular. Males also lack the blue belly patches that are apparent on male Ornate Tree Lizards. The Eastern Fence Lizard (p. 102) and Sagebrush Lizard (p. 106) also have overlapping ranges and blue belly patches on the males, but they

Adult. Near Duchesne, Uinta Basin, northeastern Utah.

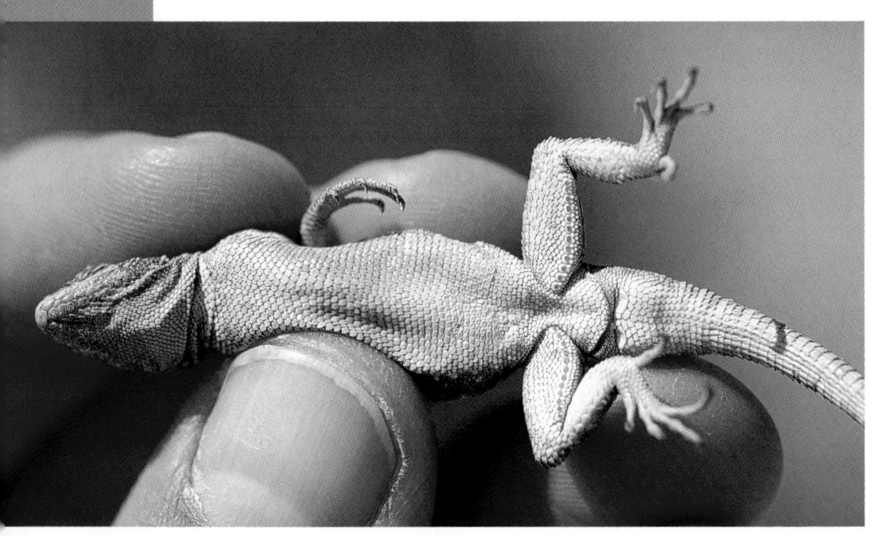

Underside of an adult male. Red Fleet State Park, Uinta Basin, northeastern Utah.

Underside of an adult female. Red Fleet State Park, Uinta Basin, northeastern Utah.

are easily recognized by their larger, prickly or rough dorsal scales and the lack of a gular fold at the throat.

DISTRIBUTION: This species has a very limited range within the area covered by this book. It occurs only in the Green River drainage of northeastern Utah, in northwestern Colorado, and in the Flaming Gorge vicinity of extreme southwestern Wyoming.

HABITAT AND BEHAVIOR: The Ornate Tree Lizard inhabits rimrocks, rocky canyons, and boulder-strewn slopes from lower deserts to the higher juniper-pinyon forests. At our northern latitude, this reptile generally does not range much above 6,500 ft (1,980 m). Despite its common name, trees are usually not the most favored basking sites. Look on the steep sides of large boulders or rock ledges, where it will often be observed

soaking up the sun's rays in a vertical position with its head pointed downward. When frightened, this lizard nimbly escapes by scurrying beyond reach, effortlessly ascending the sheer faces of rocks or climbing a tree trunk. Insects and spiders are recorded as the primary foods. Knowledge of the breeding habits for this species in our area is meager. Based upon studies conducted in Colorado, it is probable that mating takes place in May, with one to two clutches of three to eight eggs laid in late June or July, and that the hatchlings appear in late August and early September.

FIELD NOTES: 14 September 1997. I devoted nearly an hour this afternoon to photographing a Midget Faded Rattlesnake that was crawling along a rocky ledge at Red Fleet State Park, near Vernal in northeastern Utah. I was totally engrossed in getting good shots of the snake when I became aware that a creature was watching me at close range. It was a little Ornate Tree Lizard clinging to the rock face about 2 ft (61 cm) from my head. It appeared to be fairly unconcerned with my proximity, so I focused my lens on it when I was finished with the rattlesnake. Using only a 105 mm macro lens, I was able to position my camera and tripod within inches of the reptile. I marveled at how closely its coloration and pattern matched the reddish sandstone it inhabited. Just two days earlier I had photographed another Ornate Tree Lizard near Duchesne, Utah, that was on lighter, grayish-tan rocks, where it, too matched its microhabitat perfectly. Despite its common name, after several days of hiking in the area, I had not seen a single Ornate Tree Lizard on a tree. All had been on rocks. When I was through taking pictures, I slowly moved my finger toward the lizard until I gently touched the tip of its snout. Only then did it retreat into a nearby crevice.

Adult in typical sandstone habitat. Near Duchesne, Uinta Basin, northeastern Utah.

Desert Horned Lizard

Phrynosoma platyrhinos

Adult; reddish variation. Black Rock Desert, northwestern Nevada.

IDENTIFICATION: Adults are 2½–3¾ in (6–9 cm) in snout-vent length and to 5½ in (14 cm) in total length. This species conforms to the usual idea of the classic "horned toad." Despite this popular misnomer, however, it is a true reptilian lizard. **It is equipped with the distinctive, crown-like, large spines that project horizontally from the rear of the head.** The overall scalation is granular, with many small pointed scales scattered across the back. There is also a fringe of small spines along each side of the wide, very flat body and along the edges of the short tail. The Desert Horned Lizard's dorsal coloration varies from light tan and gray through shades of brown, often with a considerable amount of rusty-red speckling. There is a darker pattern on the back of gray to black spots and irregularly shaped blotches that are usually edged in white or pale gray. Each side of the neck is noticeably marked with a large, squarish, dark blotch. The undersides are white or tannish, with a sprinkling of small, dark spots. Males have enlarged postanal scales and larger femoral pores.

VARIATION: Invariably, this lizard matches the color of the soil and stones in its habitat, and Desert Horned Lizards can range from extremely pale gray or tan in areas of light sand to nearly black in places with a dark basalt substratum. Either two or three subspecies are recognized, depending upon the authority consulted. Only one occurs in the Northwest: the NORTHERN DESERT HORNED LIZARD (*P. p. platyrhinos*).

SIMILAR SPECIES: The similarly sized Greater Short-horned Lizard (p. 126) also has horizontally projecting horns, but they are considerably smaller and stubbier by comparison and are separated to the left and right

at the mid-rear of the head by a wide, deep notch (the head appears heart-shaped when viewed from above). The much smaller Pigmy Short-horned Lizard (p. 122) has tiny, nubbin-like horns that project at a nearly vertical angle from the rear of the head, in contrast to the Desert Horned Lizard's long spines that project horizontally.

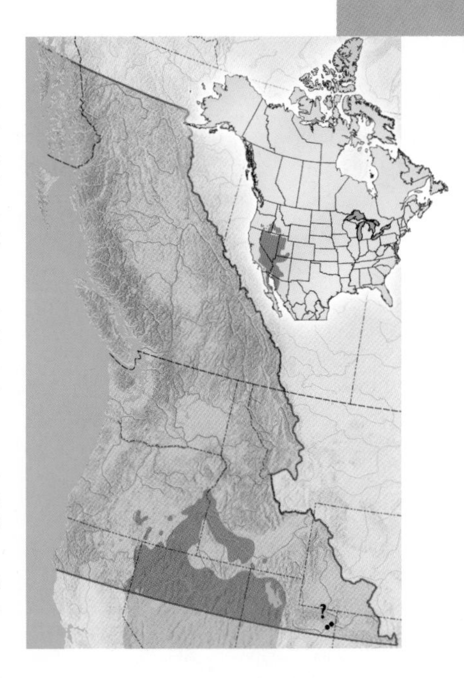

DISTRIBUTION: Within our region, this species ranges throughout most of the desert basins below 5,000 ft (1,520 m) in northern Nevada, northeastern California (Honey Lake Basin and Surprise Valley), northwestern Utah, southeastern Oregon, and southwestern Idaho. Isolated Oregon populations occur in the vicinity of several dry lakebeds in Lake County, and the Harney Basin of Harney County. Records dating from 1928 and 1942 for the Ouray-Jensen area of the Uinta Basin in northeastern Utah have not been duplicated and need verification.

HABITAT AND BEHAVIOR: The Desert Horned Lizard inhabits sandy-gravelly desert flats, brushy dune systems, and loose-soiled, open hillsides. It is most frequently encountered where desert shrubs are widely separated by sunny openings, particularly in the vicinity of dry washes. Typical plants in its habitat are Big Greasewood, Shadscale, Spiny Hopsage, and Big Sagebrush. The predatory Long-nosed Leopard Lizard often shares the same localities, sometimes preying upon smaller Desert Horned Lizards. When threatened, this lizard will initially flatten itself against the ground and remain motionless, relying on its excellent protective coloration to

Adult; pale variation. Owyhee River Canyon, southeastern Oregon.

Desert Horned Lizard

Adult; dark variation. Black Rock Desert, northwestern Nevada.

Adult. Pueblo Mountains, southeastern Oregon.

avoid detection. If this tactic fails, it will then run for cover beneath a tangled bush. When picked up, it will sometimes jerk its head backward in a weak attempt to jab one's hand with the head spines. Small rocks are often used for sunning when they are present in the habitat. The Desert Horned Lizard's pancake-shaped body allows it to easily cover itself in sand with a side-to-side, shuffling action. At sunrise it will frequently remain buried just below the surface of the sand with only the top of its broad, flat head exposed, soaking up the sun's warming rays. Ants are the primary food, supplemented with other insects and spiders. Our understanding of this species' reproductive habits in the Northwest are meager. Current information indicates that mating takes place from late April through early June, with 2 to 16 eggs deposited between June and mid-July. The hatchlings appear about two months later, usually in August to mid-September.

FIELD NOTES: 28 September 1997. After sleeping under a cold, starry sky, I crawled

from my sleeping bag at dawn and marveled at the view from my campsite above the Alvord Desert in southeastern Oregon. The crystal clarity of the air and the intensely blue sky were typical of autumn in the High Desert. Turning to gaze behind me at the lofty bulk of Steens Mountain, I noted the fresh dusting of snow at its summit and ruminated on the fact that within a couple of weeks reptiles would begin to vanish for the winter. My goal today was to visit a rattlesnake denning site. In these waning days of the season, I knew that reptiles would be out soaking up the warm sunshine at every opportunity. My assumption soon proved correct. Around 9:30 a.m., while driving one of the area's many sand-tracks that cross brushy dunes, I began seeing Desert Horned Lizards. They sunned on the sandy banks at the road margins, and I counted five over the course of just a few miles. As the morning warmed, two more scooted across the road in front of me. After spending the day checking rattlesnake hibernacula, I returned to camp in the late afternoon via the same route. The lizards were out again catching rays, some probably the same individuals I observed earlier that day. Driving roads on sunny mornings and evenings is a reliable method for finding this species and has proven successful for me on many occasions throughout the Great Basin Desert region.

Underside of an adult male. Pueblo Mountains, southeastern Oregon.

Juvenile. Pyramid Lake, northwestern Nevada.

Adult basking in evening sun along a roadside. Alvord Basin, southeastern Oregon.

Pigmy Short-horned Lizard
Phrynosoma douglasi

Adult. Fort Rock State Park, central Oregon.

IDENTIFICATION: **Adults rarely exceed 2½ in (6 cm) in snout-vent length, the average individual falling short of that size. The maximum total length of a large specimen is only about 3½ in (9 cm).** Unlike other *Phrynosoma* species, the Pigmy Short-horned Lizard's "horns" on the head are merely **small nubbins, most of which project at a nearly vertical angle**. The surface of the back has many small, pointed scales scattered among granular dorsal scales. Additionally, there is a fringe of small spines along the sides of its rather plump body and at the edges of the short tail. The overall dorsal coloration is usually gray or grayish brown, but it may be light tan, reddish, or nearly black. There is a pattern of two paired rows of dark brown to black blotches on the back that are each edged at the rear with white or pale yellow. A random light speckling is scattered throughout the dorsal pattern as well. The entire underside is white or yellowish white. Males have enlarged postanal scales and femoral pores.

VARIATION: The Pigmy Short-horned Lizard usually matches the color of the soil and rocks within its habitat, resulting in considerable diversity from one area to the next. Brick red individuals occur in some localities with red volcanic sand and pebbles, while pale tan to nearly white populations inhabit the light-colored sands of dry lake beds in the Christmas Valley area of Oregon's Lake County. There are also differences within populations that exist along the crest of the Cascade Mountains in Oregon and California. Specimens from those locations tend to be even more dwarfed and exhibit different behavioral characteristics.

SIMILAR SPECIES: Its bigger relatives, the Greater Short-horned Lizard (p. 126) and Desert Horned Lizard (p. 118) differ in having larger, well-defined horns that project horizontally at the rear of their heads. In contrast, the Pigmy Short-horned Lizard's tiny horns project almost vertically. Additionally, the Greater Short-horned Lizard has a noticeably

wide, deep notch at the mid-rear of the head, separating the horns left and right and giving its head a heart-shaped appearance when viewed from above.

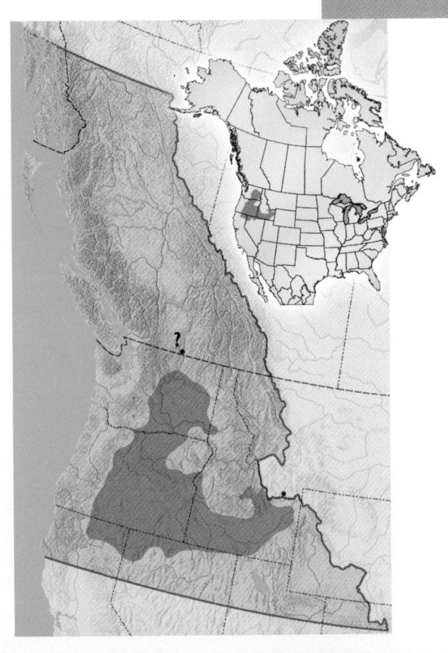

DISTRIBUTION: This uniquely Northwestern species ranges across a good portion of the dry interior plateau country and mountains of northeastern California, northwestern Nevada, southern Idaho, southeastern and central Oregon, and southeastern and central Washington. It has also been found in northern Idaho in the Snake River drainage near Lewiston. Two Pigmy Short-horned Lizards were captured in 1910 near Osoyoos, in the Okanagan Valley of south-central British Columbia. This record has never been duplicated and is in need of verification. A single specimen from barely east of the Continental Divide was collected in 1936 in extreme southwestern Montana's Centennial Valley, Beaverhead County. Other populations may exist elsewhere in western

Adult. Sand Mountain, Cascade Mountains, Oregon.

Pigmy Short-horned Lizard

Adult. Near Bend, central Oregon.

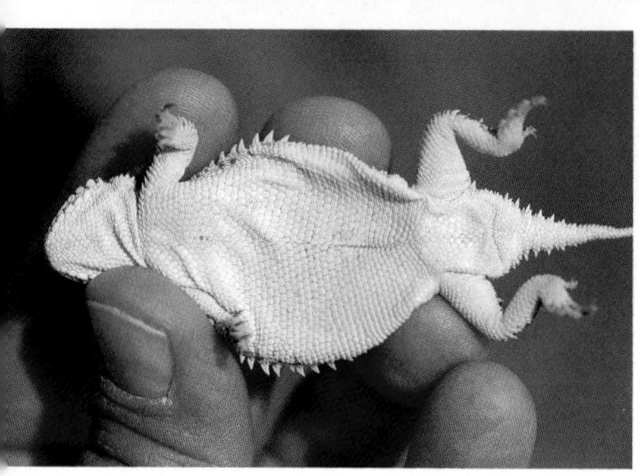

Underside of an adult female. Fort Rock State Park, central Oregon.

Montana. Elevations of at least 6,000 ft (1,830 m) are attained on some Northwestern mountain ranges.

HABITAT AND BEHAVIOR: Because of its greater tolerance to cold climates, the Pigmy Short-horned Lizard has a much wider distribution in the Northwest than our other two horned lizard species. It can be found in open sagebrush rangelands and bunchgrass plains, as well as sunny clearings in juniper and pine woodlands. In the Cascade Mountains of Oregon and northern California, it occurs in volcanic pumice sand openings in pine forests to around 5,500–6,000 ft (1,680–1,830 m) in elevation and possibly higher. There, the Pigmy Short-horned Lizard has been recorded at such places as Oregon's Santiam Pass area and Crater Lake National Park, and on the eastern slopes of California's Mt. Shasta. It appears to be absent from the higher Cascades in Washington and is restricted to lower elevations east of this mountain range in that state. The Pigmy Short-horned Lizard does not inhabit the arid salt scrub ecosystems of the northern Great Basin Desert, so it is seldom, if ever, encountered sharing habitat with the Desert Horned Lizard. Whether in open sagebrush country or forested mountains, these little reptiles usually will be found in sandy places where there are scattered bushes for cover. Occasionally, though, it will be seen on packed "hardpan" or in areas with a mixed gravelly-stony substratum. However, there will invariably be some type of loose soil nearby. Look for Pigmy Short-horned Lizards around anthills, where they seek their favorite food. To a lesser degree, other insects and their larvae are also eaten. As an adaptation to the limited warm season of higher elevations and northerly latitudes, these lizards are live-bearing, with a relatively short reproductive cycle. Mating takes place immediately after emergence from hibernation, which can range from late March through June, depending upon the elevation. From 3 to 15 young (smaller numbers in the high Cascades) are born between mid-July and mid-September.

FIELD NOTES: 19 September 2000. It was one of those grand days in Oregon's Cascade Mountains that an outdoor enthusiast can never get enough of. When long-time friend and fellow naturalist Jim Anderson and I arrived at the trailhead on Sand Mountain in the Santiam Pass vicinity, the afternoon was mostly sunny. Puffy white clouds floated through a deep blue sky, and the air had a bracing, early autumn freshness. After a short hike to the top where there is a fire lookout tower, we were treated to spectacular views of snowcapped peaks along the Cascade crest. I had wanted to find a Pigmy Short-horned Lizard in the high Cascades for many years, but my several attempts met with failure. Most people associate horned lizards with hot desert flats and cactus, not a 5,500-ft (1,680-m) summit amidst a stunted montane forest of Whitebark Pine, Subalpine Fir, and Noble Fir. As Jim put it, "These little critters must spend most of their lives in hibernation. They're snow lizards!" We immediately began searching sun-warmed openings of pumice sand. Luck was with us, for within just a few minutes I captured a female Pigmy Short-horned Lizard. While we photographed the reptile, Jim told me of a remembrance from the 1950s. He was fairly certain that he had seen a Pigmy Short-horned Lizard on top of Tam McArthur Rim in the adjoining Three Sisters Wilderness at an elevation of slightly over 7,000 ft (2,130 m), just below timberline. Also, reliable people had told him of observing them at similar elevations in the nearby Mt. Jefferson Wilderness. These tiny horned lizards of the alpine world are definitely nonconformists among the heat-loving reptiles.

Juvenile. Fort Rock State Park, central Oregon.

Adult burying in sand. Sand Mountain, Cascade Mountains, Oregon.

Naturalist Jim Anderson holding an adult Pigmy Short-horned Lizard in its high-elevation habitat on the crest of the Cascade Mountains, Oregon.

Greater Short-horned Lizard

Phrynosoma hernandesi

Adult. Ouray National Wildlife Refuge, Uinta Basin, northeastern Utah.

OTHER NAMES: Mountain Short-horned Lizard.

IDENTIFICATION: Adults are 2–4¼ in (5–11 cm) in snout-vent length and to 5¾ in (14 cm) in total length. **This relatively large and stocky horned lizard has stubby "horns" that project almost horizontally from the rear of the head. A noticeably wide, deep notch separates the right and left horns at the back of the head, creating a heart-shaped appearance when viewed from above.** The upper body has many small, pointed scales that are scattered among the otherwise granular dorsal scales. There is also a fringe of small spines along each side of the rather wide, rounded body and at the edges of the short tail. The overall dorsal coloration may range from tan, grayish brown, reddish brown, or yellowish to pale gray. There are usually two paired rows of dark brown blotches on the back that are often edged in white, along with a sprinkling of light spots throughout the dorsal pattern. These markings can be quite faint in some populations. The undersides are white or yellowish white, sometimes with a grayish suffusion or dark speckling. Males have enlarged postanal scales and slightly larger femoral pores.

VARIATION: Because this species generally matches its surrounding habitat, there is a high degree of variability in color and pattern throughout its wide range in the intermountain West. In our region, individuals from northeastern Utah, northwestern Colorado, and southwestern Wyoming generally are more brightly patterned, often with considerable reddish coloration on the body and head. Conversely, populations west of the Wasatch Mountains in northwestern Utah and in northeastern Nevada tend to be pale

gray, with very faint dorsal markings. Formerly considered to be a subspecies of the Pigmy Short-horned Lizard, full species status was recently given to the Greater Short-horned Lizard. In light of this revision, further studies are needed to determine if some of the geographic variations of this reptile constitute valid subspecies.

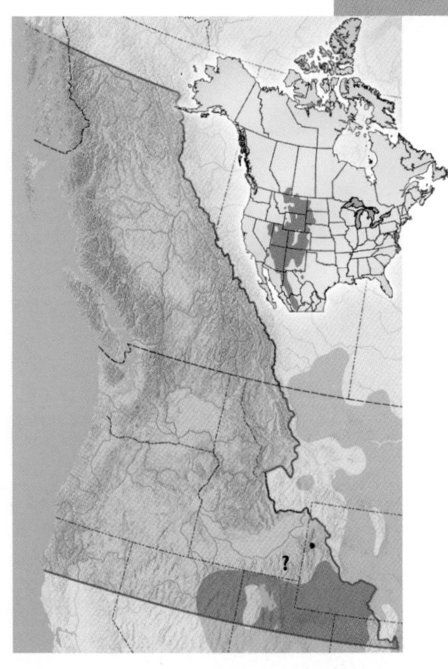

SIMILAR SPECIES: The smaller Pigmy Short-horned Lizard (p. 122) lacks the heart-shaped, deeply rear-notched head and has tiny, nubbin-like horns that project nearly vertically. The Desert Horned Lizard (p. 118) has horns that similarly project horizontally, but its horns differ in being quite large and arranged like a crown of long spines at the rear of the head with no wide, deep notch separating them left and right.

DISTRIBUTION: The Greater Short-horned Lizard has a limited range within the area covered by this guide. It is found in the Green River drainage of northeastern Utah, northwestern Colorado, and southwestern Wyoming. This reptile also inhabits the Great Salt Lake Basin of northwestern Utah (exclusive of barren playas) and the northeastern corner of Nevada. A single specimen has been collected in the upper Snake River drainage at Hoeback Canyon, Teton County, Wyoming. It may be present in the extreme southeastern corner of Idaho, but there are currently no records to confirm this distribution. Despite

Adult, showing the wide notch that gives the head a heart-shaped appearance. Ouray National Wildlife Refuge, Uinta Basin, northeastern Utah.

Underside of an adult male. Ouray National Wildlife Refuge, Uinta Basin, northeastern Utah.

Adult inflating its body with air. Ouray National Wildlife Refuge, Uinta Basin, northeastern Utah.

our northern latitude, the Greater Short-horned Lizard occurs at quite high elevations, having been recorded at nearly 8,000 ft (2,440 m) in the Uinta Mountains of northwestern Colorado.

HABITAT AND BEHAVIOR: This species occupies a variety of ecosystems, including semi-arid deserts, sagebrush steppe plains, grasslands, or open juniper and pine woodlands in the mountains. Although sandy areas are usually present within the habitat, this lizard can sometimes be found on packed soils or in places with a mixed stony, gravelly, and loose-soil composition. Search for the Greater Short-horned Lizard where the shrub and grass cover is sparse, especially around anthills. Ants are the primary food, but other insects and spiders are also eaten. When captured, it will usually inflate its body with air,

giving the appearance of a rotund, spiny balloon. Very little is known of the breeding habits for this lizard in our region. Based upon studies made at other localities within its distribution in the West, it is probable that mating takes place in April and May, with 5 to 36 young live-born between late July and early September.

FIELD NOTES: 27 August 1998, 3:00 p.m. While visiting Ouray National Wildlife Refuge in northeastern Utah, I drove to the summit of a level-topped bench that offered a good view of the Green River below. Walking off the road into a habitat of various species of bunchgrasses mixed with Shadscale, Gray Rabbitbrush, and Plains Prickly Pear Cactus, I immediately found a Greater Short-horned Lizard. Had it not moved, I wouldn't have noticed the reptile—its dorsal coloration and pattern afforded it perfect camouflage on the surrounding sandy soil that had a sprinkling of multicolored pebbles and small rocks. The lizard was easily captured for close examination of its puffed-up, fat body. As a native Oregonian, I am quite familiar with the tiny Pigmy Short-horned Lizard that inhabits my home state. Although the two species are closely related and share many similar characteristics (until recently they were considered to be subspecific variations of one species), the specimen in my hand seemed gigantic in comparison. In fact, I estimated it to be about the same size as, or slightly larger than, its other relative, the Desert Horned Lizard. However, there was no mistaking the stubby horns and deep notch at the rear of the head. After taking a few photos, I released the lizard and marveled again as it visually merged into the habitat.

Greater Short-horned Lizards can blend with their habitat. Ouray National Wildlife Refuge, northeastern Utah.

Western Skink

Eumeces skiltonianus

Young adult. Near McMinnville, Willamette Valley, northwestern Oregon.

IDENTIFICATION: Adults are 2⅛–3¼ in (5–8 cm) in snout-vent length and to about 7½ in (19 cm) in total length. The Western Skink is commonly called the "blue-tailed lizard," owing to the fact that **younger individuals of this species have brilliantly blue tails. The dorsal coloration and pattern of the body consist of rich brown, black, and golden-yellow or cream lengthwise stripes that extend from the nose to the upper tail.** As these reptiles age, the dorsal stripes and the blue of the tail become progressively more dull. In older adults, the tail often becomes a plain grayish brown. The belly is light gray to cream, with faint greenish-blue mottling. **The entire body is covered with shiny, smooth, rounded scales.** During the breeding season, the chin and sides of the head of adult males becomes suffused with reddish orange. Females sometimes have this coloration, though less intensely.

VARIATION: Three subspecies are currently recognized. The two that are native to the Northwest are similar in appearance. The SKILTON SKINK (*E. s. skiltonianus*) occurs in northern California, northwestern Nevada, most of Oregon, Washington east of the Cascades, northern Idaho, and south-central British Columbia. It has a dark-edged brown mid-dorsal stripe and light, narrow side stripes (at mid-body they are less than half the width of the mid-dorsal stripe). The GREAT BASIN SKINK (*E. s. utahensis*) occurs in north-central and eastern Nevada, northwestern Utah, southern Idaho, and possibly southeastern Oregon. Its mid-dorsal brown stripe does not have a dark edge, and the light side stripes are wider (at

mid-body they are half or more the width of the mid-dorsal stripe).

SIMILAR SPECIES: The juvenile Western Whiptail (p. 134) has a pale blue to bluish-gray tail and a dorsal pattern of lengthwise stripes, but it can be differentiated by its granular dorsal scales and a gular fold on the throat. The Western Skink's dorsal scales are considerably larger and smoother by comparison, and the tail is a much brighter blue.

DISTRIBUTION: This attractive lizard is found throughout most of the Northwest and is absent only in the highest mountains, damp coastal forests, and open arid areas. Its range extends northward from California through western Oregon (except along the coast north of Curry County), and eastward through the drier portions of the Columbia River Gorge. East of the Cascade Mountains, its distribution includes northeastern California, northern Nevada, northwestern Utah, central and eastern Oregon, Idaho, western Montana, central and eastern Washington, and south-central British Columbia. There is a past sight record for Vancouver Island near Courtenay that needs confirmation.

HABITAT AND BEHAVIOR: West of the Cascades, the Western Skink inhabits dry

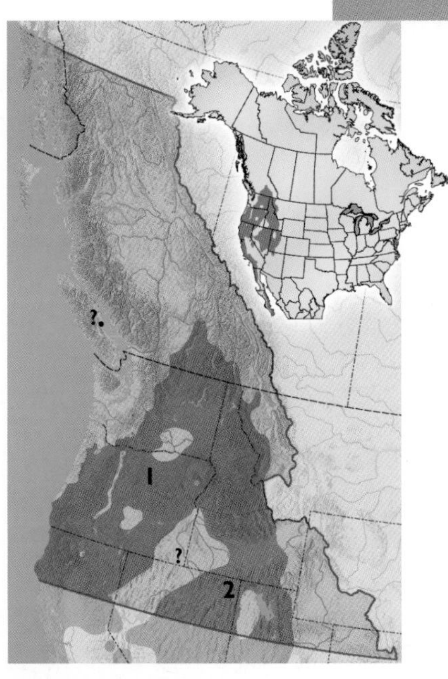

SUBSPECIES

- **1** Skilton Skink *E. s. skiltonianus*
- **2** Great Basin Skink *E. s. utahensis*
- Species range outside the Pacific Northwest

Older adult with a graying tail. Near McMinnville, Willamette Valley, northwestern Oregon.

Juvenile. Near McMinnville, Willamette Valley, northwestern Oregon.

Juvenile. Near McMinnville, Willamette Valley, northwestern Oregon.

oak woodlands, avoiding dense fir forests and damp bottomlands. To the east of this dividing range, it is found in a variety of habitats ranging from the rocky canyons of the juniper-sage rimrock country upward to mountain pine forests. This reptile sometimes occurs at elevations of nearly 7,000 ft (2,130 m) along rocky southern exposures in mountain canyons. In all areas, it has a preference for sunny, grassy openings where there are scattered rocks, logs, and leaf litter for cover. Although active during the day, it is seldom seen in the open. Look for it under flat stones and pieces of wood and bark that are resting directly on soil, where it excavates its own burrows. This little lizard is also

often common under boards around abandoned homesteads. The Western Skink is the only species of reptile native to the Northwest that will stay with its eggs and guard them until they hatch. Two to 10 eggs are deposited in a cavity under a rock or log in June or July, with hatching taking place in late July or August. Juveniles have particularly vivid blue tails which are easily broken off. The detached tail twitches violently, attracting a predator's attention while the skink escapes to grow a new tail. The food of this species consists primarily of insects and spiders, although earthworms have been eaten by captive specimens. Adults will sometimes eat young individuals of their own species.

FIELD NOTES: 20 October 1997, 4:00 p.m. After hiking all afternoon in one of the many side canyons of the eastern Columbia River Gorge in Washington, I sat down for a rest and snack. Even though the sun was out on this late autumn day, temperatures were only in the high 60s. I had visited several rocky denning sites where I had often seen rattlesnakes on past excursions, and even the rare California Mountain Kingsnake on occasion. This would probably be among the last warm days for this year and I'd hoped to get some photos of these snakes before winter arrived. However, not one reptile had appeared. As I munched some nuts and dried fruit, I sighed and resigned myself to the fact that this would probably be my last herping trip until next spring. Leaning against an oak tree and listening to the slight breeze in the branches, I suddenly became aware of another sound. It was a faint rustling in a mound of dead oak leaves nearly at my feet. As I watched, the leaves kept moving while the hidden creature progressed in halting, jerky movements. Shortly, a little head appeared, followed by the shiny, striped body and blue tail of a small Western Skink. Typical of this secretive species, it was out foraging for food in the fading sunshine, but staying undercover. Carefully, I made a move for my camera, but it saw me and, snake-like, undulated away on its short legs. Now satisfied with my hike, I got up and started for the trailhead.

Underside of an adult. Trinity River drainage, northwestern California.

Adult in typical habitat for northwestern Oregon: a hillside with rock outcrops in grassy Oregon White Oak savanna. Muddy Valley, in the foothills near McMinnville, northwestern Oregon.

Western Whiptail

Cnemidophorus tigris

Adult Great Basin Whiptail. Pueblo Mountains, southeastern Oregon.

IDENTIFICATION: A relatively large lizard, adults are 2½–4½ in (6–11 cm) in snout-vent length and to 13 in (33 cm) in total length. The tail is long (twice the length of the body or more) and the body elongated, with a narrow, pointed head and hind legs that are larger than the front legs. This lizard is the only species native to the Northwest that has a combination of **small, granular scales on the upper parts and sides of the body, and, conversely, quite large, squarish scales on the belly and chest**. The dorsal pattern is one of black or dark grayish-brown, marbled spots and blotches that are separated by light, lengthwise stripes along the back. These markings become faded and indistinct toward the rear of the body and on the tail. The ground color may be brown, tan, or gray. Ventrally, the coloration is light gray or bluish gray to tan, with dark spots on the chest and abdominal areas. There is a gular fold on the throat, and the tongue is distinctively long and forked. Males have slightly enlarged femoral pores. Juveniles display a pattern of lengthwise golden-yellow and black stripes with a pale blue to bluish-gray tail.

VARIATION: A number of subspecies have been described, with three found in the region covered by this book. The GREAT BASIN WHIPTAIL (*C. t. tigris*) has four rather dull, light grayish-brown stripes along the back, dark markings on the sides of the body that often form vertical bars, and a pale gray throat with obscure darker spots. It occurs in southeastern Oregon, southwestern Idaho, northwestern Utah, northern Nevada, and northeastern California. The CALIFORNIA WHIPTAIL (*C. t. mundus*) has eight light coppery-brown stripes along the back, dark markings on the sides of the body that usually do not form vertical bars, and large, distinct, black spots on a light throat. It occurs in the Sacramento River drainage of northern

California. The PAINTED DESERT WHIPTAIL (*C. t. septentrionalis*) has six light golden-yellow stripes along the back, dark markings on the sides that usually do not form vertical bars, and small, distinct, black spots on a light throat. It occurs in northeastern Utah and northwestern Colorado.

SIMILAR SPECIES: The Western Skink (p. 130) has much the same coloration and pattern as a juvenile Western Whiptail, but it differs in having relatively large, shiny, rounded dorsal scales, a more vivid blue tail, and no gular fold on the throat. The introduced Plateau Striped Whiptail (p. 251) has granular dorsal scalation and large, squarish ventral scales, like our native Western Whiptail, but the lengthwise stripes along its back are much more distinct, and the belly is plain white with no dark spotting. Additionally, it is found at only one limited location in central Oregon.

DISTRIBUTION: Within our region, this lizard ranges throughout much of northern Nevada, northeastern California (Honey Lake Basin and Surprise Valley), northwestern Utah, southwestern Idaho, and southeastern Oregon. It also occurs in the Sacramento River drainage of northern California, and in the Green River drainage of northeastern Utah and adjacent northwestern Colorado. Disjunct populations exist in

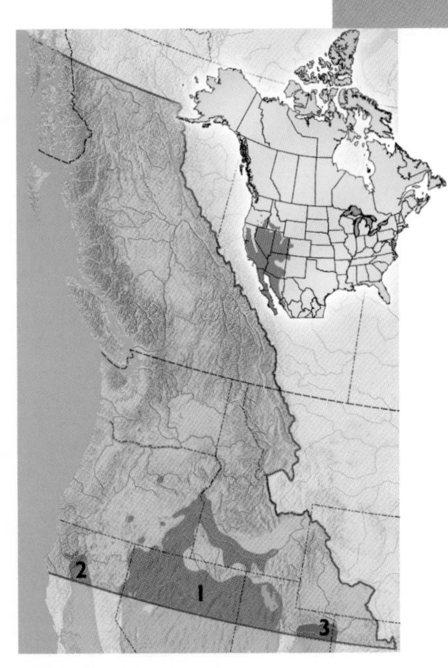

SUBSPECIES

1 Great Basin Whiptail *C. t. tigris*
2 California Whiptail *C. t. mundus*
3 Painted Desert Whiptail *C. t. septentrionalis*
Species range outside the Pacific Northwest

Adult Great Basin Whiptail with unusually vivid markings. Pyramid Lake, northwestern Nevada.

Adult California Whiptail. The Nature Conservancy's Dye Creek Preserve, near Red Bluff, Sacramento Valley, northern California.

Adult Painted Desert Whiptail. Dinosaur National Monument, northeastern Utah.

Oregon at Diamond Craters in the Harney Basin, near Crump Lake in the Warner Valley, and at John Day Fossil Beds National Monument, Sheep Rock Unit. During September 1999, I observed a juvenile Western Whiptail on the southeastern side of Summer Lake, Lake County, Oregon. In the Northwest, it inhabits elevations as low as 500 ft (150 m) near Red Bluff in California, to as high as nearly 6,000 ft (1,830 m) at Dinosaur National Monument in Colorado.

HABITAT AND BEHAVIOR: Over its wide range in the West this reptile occupies a variety of ecosystems. Primarily a desert dweller within most parts of its distribution in our region, the Western Whiptail is found on arid flats and hillsides in northern Great Basin shrub communities (saltbush, greasewood, and sagebrush species). At John Day Fossil Beds National Monument in eastern Oregon, however, and in the Uinta Basin of northeastern Utah and adjoining northwestern Colorado, it occurs in semi-arid sagebrush and juniper woodland associations. In the Sacramento Valley of northern California it inhabits open Blue Oak savannas and the surrounding chaparral country, with its mixture of various species of manzanita, buckbrush, madrone, oak, and pine. There, it is often seen along riparian areas in dry, sandy-gravelly washes where clumps of small willow and brush offer cover. No matter what ecosystem, the Western Whiptail usually favors areas that are relatively free of thick

grass and forbs which would impede foraging and running. The cooler morning hours are generally used for activity in the summer. This lizard occasionally produces a squeaking sound when captured. Foods taken include insects and their larvae, spiders, scorpions, and occasionally other lizards. It will use its front feet to dig for buried insects and sometimes use its pointed nose to turn over pieces of wood to get termites. Oddly enough, there are at least two accounts of this lizard eating such human food items as bread crumbs and discarded cooked macaroni at a campsite. Little information exists on reproduction in the Northwest. An Idaho study indicated that mating takes place in early June, with one to four eggs being laid in a burrow in late June or early July. Hatching takes place during early to mid-August.

FIELD NOTES: 8 July 1998, 2:30 p.m. I spent a very hot hour looking for lizards at Bruneau Dunes State Park in southwestern Idaho. The temperature was in the mid-90s, and I hadn't expected to find much. Just when I was ready to give up and return to my air-conditioned field vehicle, however, I noticed a Western Whiptail in some bushes at the edge of a big sand dune. I was surprised to see it, because this lizard often retreats underground to enter an inactive state of estivation (summer hibernation) during extended periods of hot weather. Typical for this alert species, it was prowling along with jerky movements, much head-bobbing, tongue flicking, and alternately lifting one fore foot and then the other off the ground. I readied my lizard noose, because I needed photos of the belly markings on a hand-held specimen. Spotting me as soon as I began stalking it, the wary reptile shot off at top speed, with its long tail held high off the ground. After taking cover behind a sagebrush it resumed its cautious movements, sneaking from bush to bush, while keeping a watchful eye on me. Eventually, I managed to get close enough to try noosing it. As so often happens with this species, though, its head slipped free when I tried to jerk the noose tight. There is little difference in width between the head and neck on all species of whiptails, and they are frustratingly difficult to capture.

Underside of an adult Great Basin Whiptail. Black Rock Desert, northwestern Nevada.

Juvenile Great Basin Whiptail. Pyramid Lake, northwestern Nevada.

Southern Alligator Lizard

Elgaria multicarinata

Adult Oregon Alligator Lizard. Near Wren, Willamette Valley, northwestern Oregon.

IDENTIFICATION: Adults are large, ranging from 2¾ in (7 cm) to nearly 6 in (15 cm) in snout-vent length and to nearly 16 in (41 cm) in total length. **The tail is usually twice the length of the body**. This lizard somewhat resembles a miniature alligator, with its large head, elongated body, and short legs. Dorsally, the pattern and coloration consists of dark, narrow, wavy crossbars on brown or olive-gray, sometimes with a reddish, greenish, or yellowish hue. The sides of the body are darker (often with a brick red or cinnamon tinge) and marked with thin, black, vertical bars that are flecked with white. The scales are large and squarish and arranged in lengthwise rows, both dorsally and ventrally. **A faint, dark line extends down the middle of each scale row on the light gray belly. A distinctive fold of skin along each side of the body reveals small, granular scales within it when spread apart. The eyes are yellow**. Males can be differentiated by their larger, broad, more triangular-shaped heads and longer tails. Juveniles have a wide, lengthwise band of copper-tan down the back and dark brown on the sides of the body.

VARIATION: Five subspecies have been described, with two found in the Northwest. There is a zone of intergradation between these two subspecies in northern California, where some individuals show characteristics of both races. The CALIFORNIA ALLIGATOR LIZARD (*E. m. multicarinata*) has a bold, well-defined dorsal pattern of wide, brick red or rusty-orange crossbands interspersed between narrower, dark crossbars and usually has dark mottling on top of the head. It barely enters our region in the Sacramento Valley of northern California. The OREGON ALLIGATOR LIZARD (*E. m. scincicauda*) usually has a uniformly brown, olive-gray, or subdued reddish-brown

back with narrow dark crossbars and no dark mottling on top of the head. It occurs north of California's Sacramento Valley, through Oregon and Washington.

SIMILAR SPECIES: The Northern Alligator Lizard (p. 142) differs in having brown (not yellow) eyes.

DISTRIBUTION: This lizard is common throughout all but the most densely forested areas of both the coastal and inland portions of northern California and southwestern Oregon. In western Oregon north of Curry, Josephine, and Jackson counties, however, it is primarily confined to the interior Umpqua and Willamette valleys. It penetrates east of the Cascade Mountains via the Columbia River Gorge into north-central Oregon and south-central Washington. This species also ranges up the Klamath River drainage to the eastern flanks of the Cascades in the southwestern corner of Oregon's Klamath County and California's northeastern portion of Siskiyou County. Relatively low elevations are inhabited, rarely exceeding 3,000 ft (910 m).

HABITAT AND BEHAVIOR: West of the Cascade Mountains, the Southern Alligator Lizard is closely associated with the oak woodlands and savannas of the foothills. In the drier Klamath Mountain region of southwestern Oregon and northwestern

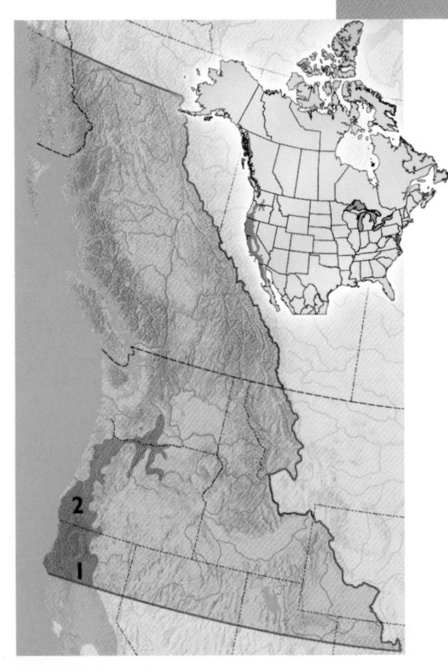

SUBSPECIES

1 California Alligator Lizard *E. m. multicarinata*

Area of intergradation

2 Oregon Alligator Lizard *E. m. scincicauda*

Species range outside the Pacific Northwest

Adult California Alligator Lizard. Redding, Sacramento Valley, northern California.

Southern Alligator Lizard

Adult Oregon Alligator Lizard, showing its yellow eye. Columbia River Gorge National Scenic Area, near Rowena, Oregon.

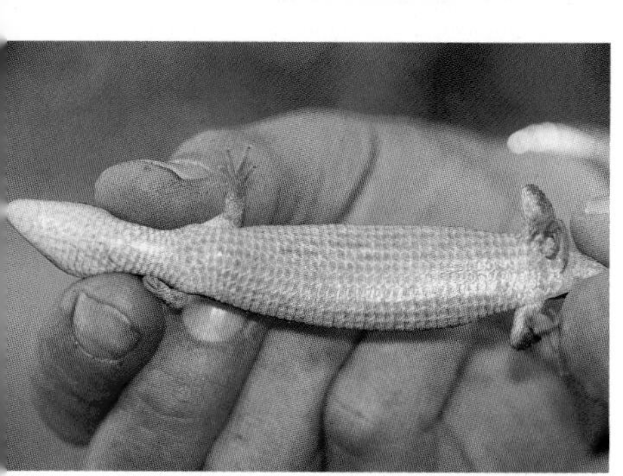

Underside of an adult Oregon Alligator Lizard, showing faint darker lines extending down the middle of the scale rows. Columbia River Gorge National Scenic Area, near Rowena, Oregon.

California it ranges higher into the more open sections of pine forests and mixed oak-pine and brushy chaparral country. East of the Cascades, in the Columbia River Basin of south-central Washington and north-central Oregon, it occurs among a growth of sagebrush, bitterbrush, juniper, and pine. In these semi-arid locations, this lizard is often somewhat confined to the riparian areas of canyon bottoms, particularly where there are patches of jumbled rock talus for cover. Although frequently observed out foraging, it can also be found under logs, bark, and stones. With its short legs and elongated body, it uses an undulating, snake-like motion when escaping. Occasionally, this species will be encountered as it climbs in bushes, using its prehensile tail as a climbing aid by wrapping it around branches. Southern Alligator Lizards feed on insects, spiders, centipedes, scorpions, slugs, snails, earthworms, and smaller lizards. Reportedly, young mice are also eaten, and bird nests are sometimes raided for eggs and nestlings. At our northerly latitude, 8 to 14 eggs are laid in a burrow or crevice during summer (probably July or August) and hatch in September or early October.

FIELD NOTES: 30 April 1998. As I worked my way along the summit of the southern rim of Lower Table Rock in Oregon's Rogue River Valley, the afternoon sun was

momentarily shining. It had been a cold, rainy spring, and I could see more clouds advancing on the western horizon. Managed by The Nature Conservancy, this preserve is usually a great place to find reptiles, but I'd have to work quickly today before temperatures cooled again. Before long, I turned over a flat, sun-warmed rock and discovered a big, male Southern Alligator Lizard. It immediately gaped its jaws in a threatening manner and glared at me with its little, yellow eyes. The message was clear: "Leave me alone or you'll be sorry!" I made a fast grab for the lizard's neck and avoided a painful bite. Writhing about, it began defecating foul-smelling feces all over my hand, a typical defense reaction for this species. Undeterred, I readied my camera with my free hand, positioned the lizard on a rock, slowly released it, and hoped that it would stay put for a moment or two. As I squeezed off two quick shots before the lizard dove into a crevice, the wind picked up and I could feel the threatening clouds over my shoulder. I made a hasty retreat back down the 2-mi (3-km) trail to the parking lot and arrived drenched by cold rain.

Juvenile Oregon Alligator Lizard. Columbia River Gorge National Scenic Area, near Lyle, Washington.

Southern Alligator Lizards rarely fail to bite vigorously when captured. Near The Dalles, north-central Oregon.

Northern Alligator Lizard

Elgaria coerulea

Adult Northwestern Alligator Lizard. Near Victoria, Vancouver Island, British Columbia.

IDENTIFICATION: Adults are 2¾–5¼ in (7–13 cm) in snout-vent length and to 10 in (25 cm) in total length. The body is elongated and the legs are short. Dorsally, the coloration of the back is brown, tan, olive, gray, yellow, or greenish yellow. The usually darker sides of the body are checkered with still darker squarish blotches and bars. Some populations may have a row of dark spots down the middle of the back, while others have dark crossbars. **The belly is uniformly pale gray to white. A distinctive fold of skin running along each side of the body reveals small, granular scales within it when spread apart. The eyes are brown.** Males differ in having larger, broader, triangular-shaped heads. Juveniles have a wide band of copper-tan down the back.

VARIATION: Four subspecies are recognized, two of which are found in the region covered by this book. There appears to be an area of intergradation in extreme northern California and southwestern Oregon, where individuals with characteristics of both races may be encountered. The NORTHWESTERN ALLIGATOR LIZARD (*E. c. principis*) is smaller, usually to about 4 in (10 cm) in snout-vent length, and has a dorsal coloration of brown, tan, olive, or grayish, usually with a row of dark spots down the middle of the back. The dark markings on the sides lack white flecks. It occurs throughout most of this reptile's range in the Northwest, except for the Klamath Mountains of southwestern Oregon and northern California. The SHASTA ALLIGATOR LIZARD (*E. c. shastensis*) is larger, to 5¼ in (13 cm) in snout-vent length, and generally more robust. It is variable in color and pattern: males often have bright yellow or greenish-yellow backs and gray heads in California, and the dark markings on the sides are flecked with white; females are usually more dull colored. In some parts of its range, all individuals are brown, tan, or yellowish, with dark dorsal crossbars. It occurs in the Klamath Mountains of southwestern Oregon and

northern California, and several locations in the adjoining northern Great Basin. Specimens with pale yellow backs and bold irregular black crossbars have been found in the Applegate River drainage of Oregon's Siskiyou Mountains.

SIMILAR SPECIES: The Southern Alligator Lizard (p. 138) differs in having yellow (not brown) eyes and faint, dark lines that extend down the middle (not edges) of each scale row on the belly.

DISTRIBUTION: This lizard occupies a swath of the Northwest that extends west from the Cascade Mountains to the coast, and north from California to southern British Columbia. It ranges east across northeastern Washington into the Rocky Mountain region of northern Idaho and northwestern Montana. In addition, the Northern Alligator Lizard has been recorded on the eastern portions of Vancouver Island and a number of the smaller adjacent islands. Disjunct populations that are currently thought to be of the Shasta Alligator Lizard subspecies occur east of the Cascades in some of the isolated ranges of the northern Great Basin. Records there include Hart Mountain and the Gearhart Mountain area in south-central Oregon, the Warner Mountains in northeastern California, and Badger Mountain and High Rock Canyon in northwestern Nevada.

HABITAT AND BEHAVIOR: A truly Northwest reptile, the Northern Alligator Lizard is associated primarily with coniferous forests and is the only species of lizard found on the northern Oregon and Washington coasts (although it is absent from the rain-soaked western side of the Olympic Peninsula). It ranges to higher elevations than the Southern Alligator Lizard, sometimes being found at up to 7,000 ft (2,130 m). Although it is more tolerant of cool, damp environments than most reptiles, some sunny clearings are required. Look for it around the edges of meadows, roadcuts, abandoned homesteads and sawmill sites, or where logging has opened the tree canopy. In such places, it is often fairly common under logs, bark, boards

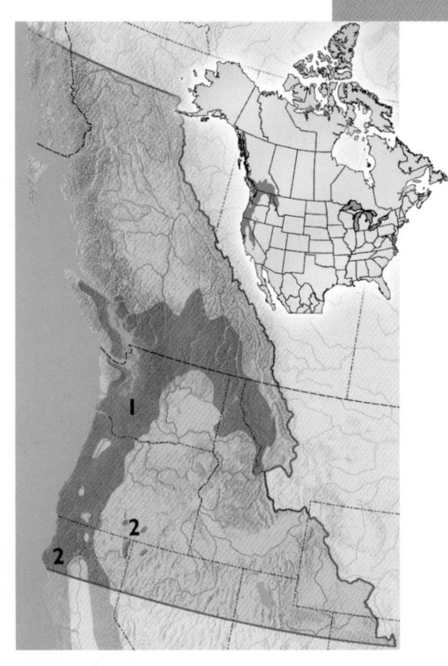

SUBSPECIES

- 1 — Northwestern Alligator Lizard *E. c. principis*
- Area of intergradation
- 2 — Shasta Alligator Lizard *E. c. shastensis*
- Species range outside the Pacific Northwest

Adult male Shasta Alligator Lizard. The Nature Conservancy's McCloud River Preserve, northern California.

Adult; probably an intergrade between the Northwestern and Shasta subspecies. Howard Prairie, southern Cascade Mountains, Oregon.

and other surface objects. Open valley floors and the oak woodlands and savannas of the foothills are avoided throughout most of its distribution. The Shasta Alligator Lizard subspecies, however, is sometimes found in drier situations. At Oregon's Hart Mountain National Antelope Refuge it occurs in relatively exposed, rocky areas of Big Sagebrush, Gray Rabbitbrush, Bitterbrush, and Western Juniper. Foods consist mainly of insects and their larvae, spiders, centipedes, millipedes, ticks, slugs, and worms. Undoubtedly as an adaptation to the shady forests of cooler northern climates, this lizard is live-bearing, with two to eight young born between July and early September.

Adult Northwestern Alligator Lizard, showing its brown eye. Chetco River drainage, southern Oregon coast.

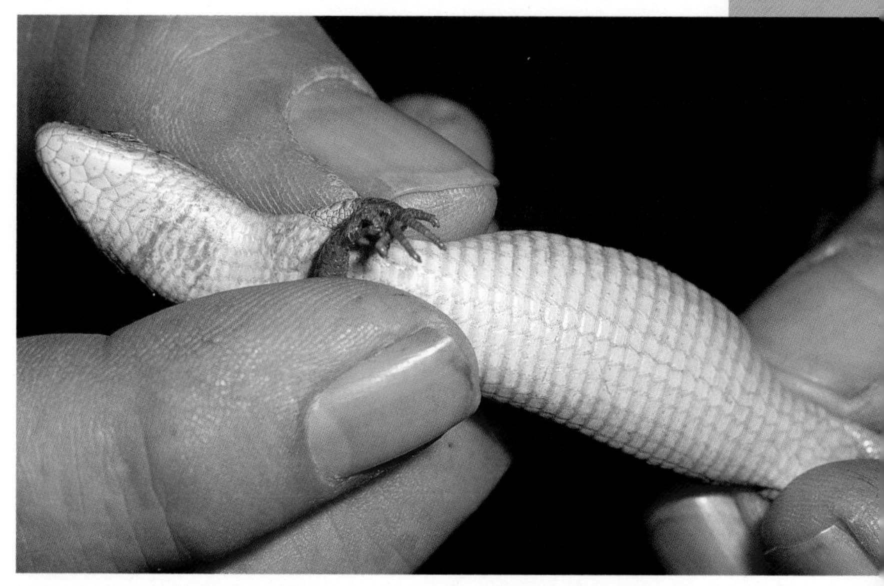

Underside of an adult Northwestern Alligator Lizard, showing the uniformly pale belly. Chetco River drainage, southern Oregon coast.

FIELD NOTES: 16 June 1998. Lee Simons, along with his son, Clayton, and nephew, Matthew, accompanied me to The Nature Conservancy's McCloud River Preserve in northern California. Lee, a local biologist, grew up in the area and is extremely knowledgeable about the native reptiles. His help was a real boon, and my excitement grew as we hiked the trail down into the beautiful river canyon with its towering firs and cedars. Lee's description of the male Shasta Alligator Lizards found there painted an intriguing picture of an exceptionally colorful reptile. I was familiar with the smaller, rather dull, olive-tan Northwestern Alligator Lizard subspecies to the north in my native Oregon. I had also seen lizards in southwestern Oregon that were presumed to be of the Shasta race, but they were plain brown with dark crossbars. Within a short time, Matthew captured our first specimen, a female that was merely brown colored. Lee, however, who had been poking about among rocks and logs a short distance away, soon approached with a big smile on his face and handed me a snake bag. Peering inside, I was startled by what I saw; a large, male Shasta Alligator Lizard that sported a bright, chartreuse-like, greenish-yellow body and a contrasting slate gray head. It was unlike any alligator lizard I'd ever seen. After the reptile was photographed and released, we had a lively discussion about this species' taxonomy as we hiked out to the trailhead. Clearly, there is need for in-depth study of the divergent populations in the Klamath region and the isolated mountain ranges of the northern Great Basin.

Juvenile Northwestern Alligator Lizard. Near Victoria, Vancouver Island, British Columbia.

The Snakes

Western Rattlesnake. Columbia River Gorge National Scenic Area, Oregon.

Rubber Boa
Charina bottae

Adult. Amity Hills, Willamette Valley, northwestern Oregon.

IDENTIFICATION: A small, **thick-bodied snake.** Adults are usually 17–27 in (43–69 cm) long, although a single 33-in (84-cm) specimen has been recorded. **The small, smooth dorsal scales and soft, loose skin give this uniformly light tan to dark brown or olive-brown snake a rubbery appearance.** It is commonly called the "two-headed snake," because its **short, blunt tail** resembles its small head. The tip of the tail has a large, solid plate that creates a protective, hard cap. Unlike all other snakes native to the Northwest, it has **no enlarged chin shields. The eyes are small and have vertical pupils.** Ventrally, the coloration may be cream, yellow, or orangish yellow, with some individuals having extensive dark mottling on the belly. As in all species of boas, there are tiny vestigial legs in the form of a dark spur on each side of the vent. Males differ in having more pronounced spurs and a relatively longer tail. Females are usually more heavy bodied, may attain a greater length, and have much smaller spurs that are rarely visible. Juveniles are light pinkish tan and look somewhat like an earthworm.

VARIATION: Among populations of Rubber Boas, there are considerable individual differences in scalation, dorsal and ventral coloration, as well as the extent or lack of dark mottling on the belly. As a result, three subspecies have been described in the past. Pending further studies, however, the two subspecies proposed for the Northwest are currently not recognized by most authorities.

SIMILAR SPECIES: An adult Racer (p. 162) has much the same coloration and is sometimes mistaken for a Rubber Boa.

Adult. Fort Rock State Park, central Oregon.

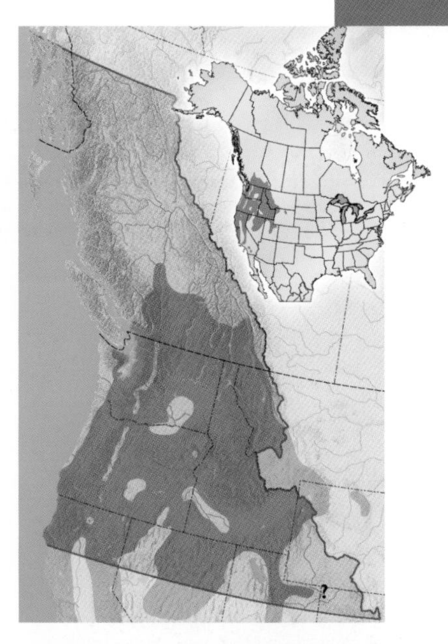

A Racer differs in having a long, pointed tail, a thinner body, and large eyes with round pupils, and it is a faster-moving snake in comparison with the sluggish Rubber Boa.

DISTRIBUTION: This reptile is found throughout most of the Northwest, usually at elevations below 7,000 ft (2,130 m). It has been recorded from as high as 8,500 ft (2,590 m) or more in the mountains of Utah and in Yellowstone National Park in Wyoming. It ranges north from California through Oregon, Washington (apparently absent from Puget Sound islands), and southern British Columbia, then eastward across Idaho, western Montana, northwestern Wyoming, and northern Utah. It is reported to occur in extreme northwestern Colorado at Dinosaur National Monument, but validation is needed. A past record for the southern end of Vancouver Island, British Columbia, is probably erroneous.

HABITAT AND BEHAVIOR: West of the Cascade Mountains, the Rubber Boa occupies a variety of habitats, including oak woodlands, grassy savannas, coniferous forests, and, in the Klamath Mountains region, a mixture of oak, pine, and brushy chaparral. Extensive sections of dense, shady fir and cedar forests without clearings are avoided. Despite an ability to be active at relatively low temperatures, it appears that sufficient sunshine is required to provide warmth for embryonic development and resulting successful reproduction. Probably for this reason, it is entirely absent from the coastal rainforests of the Northwest north of Coos County, Oregon. To the east of the Cascades,

Adult. Amity Hills, Willamette Valley, northwestern Oregon.

Rubber Boa

Variations of belly patterns on adult Rubber Boas.

Newly born Rubber Boa. Corvallis, Willamette Valley, northwestern Oregon.

this snake inhabits dry pine and juniper woods and higher montane forests, but it is usually not found in areas of treeless plains and arid deserts. Occasionally it will be found in open regions along riparian zones. One exception is at Fort Rock State Park, which is situated several miles from the nearest water or forest in the sagebrush country of Oregon's Lake County. The Rubber Boa is a secretive reptile and is primarily active at dusk and at night, although it is often seen abroad in the daytime during warm, cloudy weather. It is most commonly found in rotting stumps or under logs, bark, rocks, and other objects in grassy openings among trees, often near a streamcourse. Also look for it under boards, pieces of tin, and other litter around old, abandoned houses and outbuildings. A good burrower, this snake can sometimes be discovered hidden just below the surface of dead leaf litter. On more than one occasion, I have found congregations of Rubber Boas in old, decomposing sawdust piles. These observations were often made in the early spring, so the sawdust was probably being used for hibernation. Slow-moving and timid, a Rubber Boa never attempts to defend itself by biting. When confronted with danger it usually rolls into a ball, hides its head under its coils, and then protrudes its stubby tail as a decoy head. Occasionally, it will jab about with the tail, simulating a striking movement. In this way, a predator will be attracted to the more expendable tail while the actual head is protected from harm. A bad-smelling musk is emitted from the vent as an additional attempt to repel enemies. The preferred food is small rodents (mainly mice and shrews), particularly the young in a nest. While the snake eats the infant rodents, it uses its hard-tipped tail as a jousting weapon to club the attacking mother. The rear portions of most individuals have many scars that probably result from the use of this defensive tactic. Other occasional food items are lizards, snakes, salamanders, small chipmunks, nestling rabbits, and bats. This snake is also reported to climb trees in pursuit of nestling birds. When available, lizard and snake eggs may be consumed. As with all members of the boa families, the prey is subdued by constriction. From one to eight young are live-born from mid August (occasionally early August) through October, and sometimes as late in the season as early November.

Adult in a defensive display of hiding its head under the body coils and displaying its blunt tail. Near Brookings, southern Oregon coast.

FIELD NOTES: 6 October 1998. A sunny autumn afternoon was devoted to looking for reptiles in the Amity Hills of Yamhill County in Oregon's Willamette Valley. I had the help of friends Doug Knutsen, Richard Hoyer, Laura McMasters, and her grandson, Sebastian. We had a pleasant hike through the golden-hued oaks on the south-facing slope, finding a Southern Alligator Lizard and a Sharp-tailed Snake. Richard has conducted extensive studies on Rubber Boas for many years and is an expert on all aspects of their natural history. I never cease to be amazed by his ability to find these snakes almost anyplace, and today was no exception. When we returned to where our cars were parked, Richard produced a nice Rubber Boa from under a small log on the road bank. Sebastian placed the docile reptile around his neck, where it remained like a living necklace until being returned to its original hiding place under the log. This behavior is typical for this species, and I have often seen Rubber Boas that would coil around someone's wrist for an hour or two at a time.

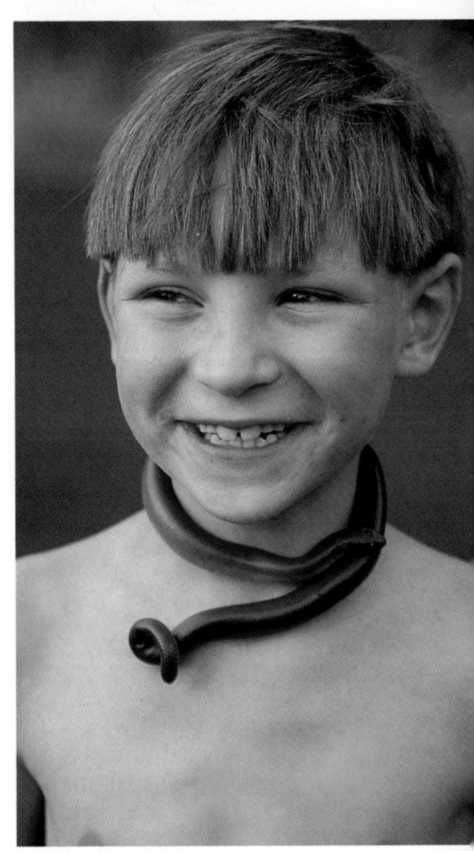

Sebastian McMasters and a Rubber Boa. Amity Hills, Willamette Valley, northwestern Oregon.

Sharp-tailed Snake
Contia tenuis

Adult; grayish-brown variation. Near Ashland, Rogue River Valley, southwestern Oregon.

IDENTIFICATION: One of the smallest species of Northwestern snakes, adults typically average only 8–12 in (20–30 cm) long. The maximum size attained in our region is probably no more than 14–15 in (36–38 cm). **A sharp spine is located at the tip of the short tail,** giving this snake its common name. The dorsal coloration is uniform reddish brown to grayish brown, usually with a narrow line of copper red along each side of the back (sometimes faint). **The pale gray to cream-colored belly is distinctively marked with a black crossbar along the upper half of each ventral scale, creating an alternating dark-and-light, ladder-like pattern.** The dorsal scales are smooth. The head is flattened, with a somewhat squared-off snout, creating a slight boxy shape. A dark mask stripe is present on each side of the head. The eyes have round pupils. Juveniles are marked with red lines that are considerably more vivid than those of adults, often with the entire back being bright copper red to orangish red.

VARIATION: It was recently discovered that there is an overlooked form of the Sharp-tailed Snake that has a significantly longer tail and may prove to be a totally new species. Preliminary studies indicate that it is larger in size, possibly reaching 17–18 in (43–46 cm). This snake is apparently confined to coniferous forests of the mountains in northern California and southwestern Oregon, reportedly to elevations of nearly 5,000 ft (1,520 m).

SIMILAR SPECIES: Small Rubber Boas (p. 148) or Racers (p. 162), which are also uniformly brown dorsally, look similar, but they lack the tail spine and dark crossbars on the belly. The Northwestern Garter Snake (p. 216) often has red dorsal stripes, and juveniles and small subadults could be mistaken for a Sharp-tailed Snake. They differ in having keeled dorsal scales.

DISTRIBUTION: The Sharp-tailed Snake occurs from valley floors to mountainous terrain in the Northwest, at elevations from just above sea level to around 4,000 ft (1,220 m). Its range is somewhat continuous in northern California and southwestern Oregon but becomes more spotty to the

north. In northwestern Oregon it is largely confined to the Willamette Valley. Most records there come from the western portions of that drainage and the adjoining foothills of the Coast Range as far north as Yamhill County. There is a reported observation from near the summit of Mary's Peak in the Coast Range of Benton County, Oregon. The only documented records for western Washington come from the Puget Sound area near Tacoma, but this snake has not been found there in recent years. Other sites from west of the Cascade Mountains are in British Columbia at the southern end of Vancouver Island and on some of the nearby Gulf Islands. This reptile is also known from several locations on the Washington side of the Columbia River Gorge in Skamania and Klickitat counties. Additionally, there are populations scattered along the eastern slopes of the Cascades. Interior localities are at Meiss Lake in northern California's Siskiyou County, the vicinities of Tygh Valley and The Dalles in Wasco County of north-central Oregon, and in central Washington in the Yakima River Canyon, Kittitas County, and the Leavenworth area, Chelan County. Two specimens were reported as having been collected in 1964 near Chase in south-central British Columbia, but updated confirmation is needed.

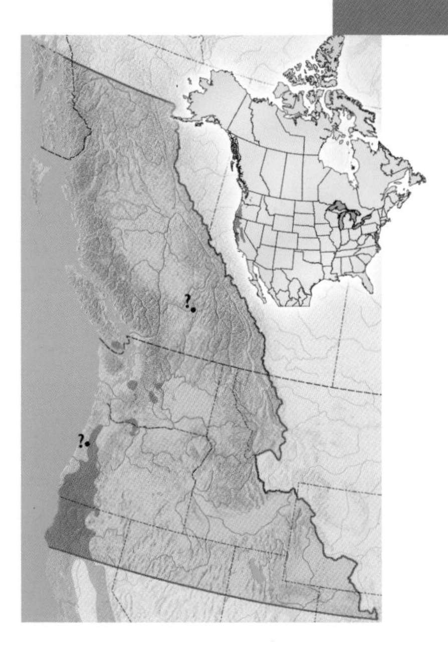

HABITAT AND BEHAVIOR: The Sharp-tailed Snake is usually associated with oak woodlands over much of its range, but it can also be found in open pine forests, dense coniferous forests, brushy chaparral country, and grassy savannas. In the dry climate east of the Cascades it is sometimes confined to the vicinity of moist riparian zones. Although this snake is often active at relatively low temperatures and has been found in shady forests of fir and cedar and coastal redwood groves in the Klamath Mountain

Adult; reddish-brown variation. Tygh Valley, north-central Oregon.

Adult. Corvallis, Willamette Valley, northwestern Oregon.

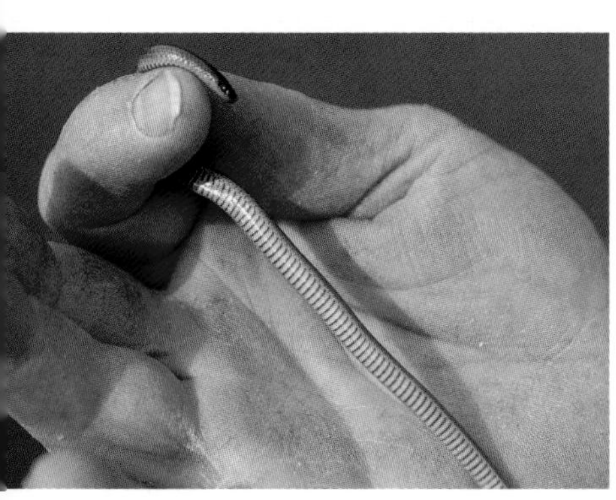

Underside of an adult, showing the dark crossbars on the ventral scales of the belly. Corvallis, Willamette Valley, northwestern Oregon.

region (long-tailed form), forest openings appear to be required in the northerly parts of its distribution. The Sharp-tailed Snake is extremely secretive, leading a largely subterranean existence or hiding under surface rocks, logs, pieces of bark, leaf litter, and boards. A slightly damp microhabitat with slugs seems to be required. Surface activity occurs mainly in the spring and autumn—the most productive time to search for this species is on a warm, humid day after rains in April or October. It is sometimes nocturnally active during the summer. Aggregations of relatively large numbers are often found together under objects at one site. Juveniles sometimes coil, hide the head, and display the dark-and-light crossbanding on their bellies, possibly mimicking toxic millipedes to deter predators. The food for this reptile consists almost entirely of small slugs, although slug and snail eggs, flatworms, wireworms, earthworms, and slender salamanders (*Batrachoseps* species) have reportedly been eaten by individuals in captivity. Initial observations indicate that the tail spine is employed as an anchoring hook in the substratum when a captured slug attempts to pull away and escape. The scant information on reproduction suggests that one to five eggs are laid in June or July (sometimes communally with other females) and hatch in late August, September, or early October.

FIELD NOTES: 25 September 1999. I spent an hour this afternoon photographing an unusual Sharp-tailed Snake in Richard Hoyer's backyard in Corvallis, Oregon. The

Tail of an adult. Corvallis, Willamette Valley, northwestern Oregon.

small reptile had been caught in Oregon's Siskiyou Mountains. Richard had made a surprising discovery while conducting an in-depth survey of this species. Several had tails that were notably longer than the others he had found, along with some differences in scalation. All came from forested mountains in southwestern Oregon and northern California. Admirably thorough, Hoyer checked hundreds of preserved specimens in institutional collections and found that there were about 80 of the long-tailed form represented. Some had been collected more than 100 years ago, but were overlooked as being different. I was guilty of the same oversight, having captured and photographed two of these Sharp-tails along the North Umpqua River in the Cascade Mountains of Oregon during 1985. I had noticed that their tails were quite long and wondered about it, but got busy with other projects and didn't delve further. When Richard told me of his observations, it jogged my memory and I dug the slides out of my files. Sure enough, they had long tails. What a discovery! A possible new species right under our noses all those years. There are still exciting new things to be learned by the student of nature who pays attention to details.

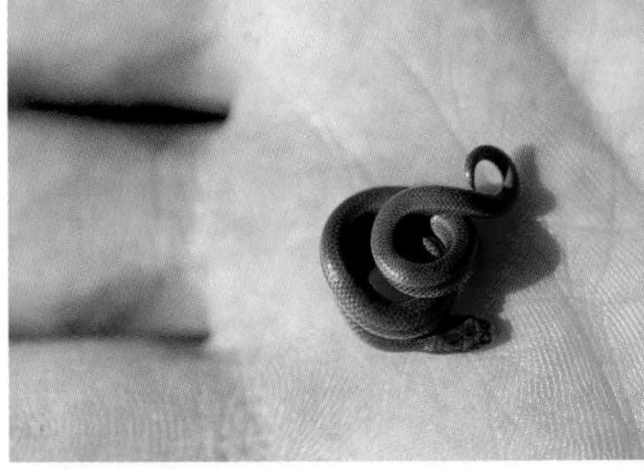

Hatchling. Saltspring Island, British Columbia.

Hatchling in a defensive coiling display. Saltspring Island, British Columbia.

The recently discovered long-tailed form of the Sharp-tailed Snake. Near Glendale, Siskiyou Mountains, southwestern Oregon.

Ring-necked Snake
Diadophis punctatus

Adult Northwestern Ring-necked Snake. Wilson Wildlife Area, near Corvallis, Willamette Valley, northwestern Oregon.

IDENTIFICATION: Adults may range from 8 in (20 cm) to 30 in (76 cm) in length, although most individuals are less than 18 in (46 cm). This snake is two-toned, with **slate gray coloration above and contrasting bright reddish orange or yellow below. A ring of matching orange or yellow encircles the neck (absent in some populations)**, from which the common name is derived. The underside of the tail is colored with a particularly intense hue of reddish orange, while the belly is usually sprinkled with black spots. The overall scalation is highly glossy, with the smooth dorsal scales exhibiting a beautiful greenish-blue iridescence. Juveniles are dark gray to nearly black dorsally.

VARIATION: Considerable divergence exists within the coast-to-coast range of this snake in North America. Some herpetologists favor considering those in the West to be a distinct species, *Diadophis amabilis*. Twelve subspecies are currently recognized in the U.S. and Canada, with three native to the area covered by this book. Recent molecular genetics studies indicate that some of the West Coast races are of questionable validity. The Northwestern Ring-necked Snake (*D. p. occidentalis*) has a wide, vivid, orange neck ring; its reddish-orange ventral coloration extends up the sides of the body onto the first $1^{1}/_{2}$ to 2 scale rows and is usually flecked with black; and its belly has a moderate to heavy scattering of black spots throughout. It occurs in northwestern California, Oregon, Washington, and western Idaho. The Coral-bellied Ring-necked Snake (*D. p. pulchellus*) is very similar to the Northwestern Ring-necked Snake, but the first reddish-orange scale rows on the sides of the body usually lack black flecks, and the belly often tends toward bright red (sometimes intense coral red to scarlet) and is only lightly sprinkled with black spots or completely devoid of such

markings. It occurs in California's Sierra Nevada, probably intergrading with the Northwestern Ring-necked Snake in the Cascade/Klamath Mountains of northern California and southern Oregon. The REGAL RING-NECKED SNAKE (*D. p. regalis*) often has an obscure or absent neck ring; the reddish-orange to yellow ventral coloration extends up the sides of the body on only one scale row and is usually flecked with black; the belly is often heavily spotted with black; and the gray dorsal coloration tends to be pale, with a distinct bluish or greenish cast. This Ring-necked Snake is generally larger in size than other subspecies. It occurs in southeastern Idaho, Utah, and eastern Nevada.

SIMILAR SPECIES: The Ring-necked Snake is easily distinguished from all other snakes in the Northwest by the uniformly slate gray dorsal color that is sharply contrasted by bright reddish-orange or yellow ventral areas.

DISTRIBUTION: This species occupies elevations from just above sea level to 2,000–3,000 ft (610–910 m) over most of its distribution in the Northwest. It has been found as high as 7,000 ft (2,130 m) in Utah. It ranges from California north through most of western Oregon, except along the coast north of Coos County. In western Washington it has been recorded only along the Columbia River in the southern portions of Cowlitz, Clark, and Skamania counties. The Ring-necked Snake penetrates east through the Columbia River Gorge into the lower Deschutes River drainage of north-central Oregon and in Washington north along the eastern slopes of the Cascades to Kittitas County. Eastward, in the intermountain Northwest, there are disjunct populations in the Snake River drainage of southeastern Washington and western Idaho and at scattered locations throughout the mountain ranges of southeastern Idaho and northern Utah. Additionally, this reptile has been found in the Granite Range of Washoe County in northwestern Nevada, and it possibly occurs in other mountains of the northern parts of that state and adjoining northeastern California. In extreme northeastern Oregon, the

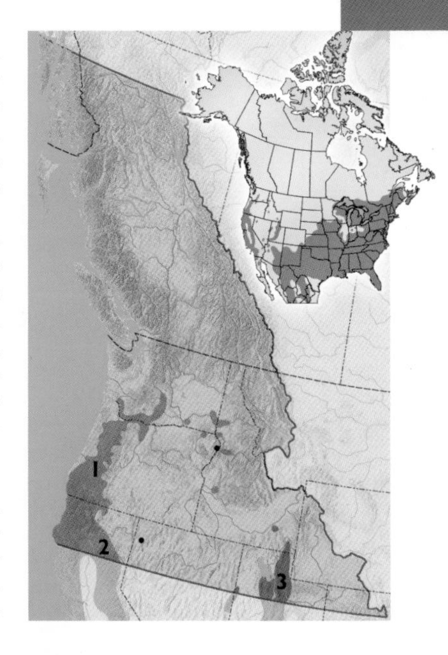

SUBSPECIES

- 1 Northwestern Ring-necked Snake *D. p. occidentalis*
- Area of intergradation
- 2 Coral-bellied Ring-necked Snake *D. p. pulchellus*
- 3 Regal Ring-necked Snake *D. p. regalis*
- Species range outside the Northwest

Adult Northwestern Ring-necked Snake. Near Lebanon, Willamette Valley, northwestern Oregon.

Ring-necked Snake

Juvenile Northwestern Ring-necked Snakes. Near McMinnville, Willamette Valley, northwestern Oregon.

Underside of an adult with a missing tail; probably an intergrade between the Northwestern and Coral-bellied subspecies. Near La Moine, upper Sacramento River drainage, northern California.

Ring-necked Snake has been recorded in Hells Canyon and along the Grande Ronde River near Troy.

HABITAT AND BEHAVIOR: In the western portions of our region the Ring-necked Snake inhabits mixed oak and conifer woodlands, brushy chaparral country, and the borders of grassy savannas. It is sometimes found on the more open valley floors in sufficiently well-drained locations. As the aridity increases east of the Cascades, it is largely confined to moist riparian areas and mountain forests. In Utah, the *regalis* subspecies often ranges into the higher zones of aspen, maple, and fir. Although occasionally encountered crawling in the open, this secretive species is usually found beneath rocks, logs, pieces of bark, boards, and leaf debris. When threatened, it emits a pungent musk from the vent, hides its head beneath the body, and coils the tail upward like a corkscrew, displaying the bright red underside. These actions presumably divert or intimidate an attacker. Foods consist of salamanders, frogs, worms, slugs, insects, lizards, and small snakes. There is evidence that this snake's saliva may be toxic and possibly acts as an aid in subduing struggling prey. Our

Adult Regal Ring-necked Snake. Near Provo, northwestern Utah.

meager information for the Northwest indicates that a single yearly clutch of 1 to 10 eggs is laid in early July and hatches in September.

FIELD NOTES: 17 April 1998. Richard Hoyer acted as my guide at the E. E. Wilson Wildlife Area north of Corvallis. It was a balmy spring day in northwestern Oregon and temperatures had reached the low 70s. While discussing the habitat preferences of the local species of snakes, I remarked that I considered the Ring-necked Snake to be a dweller of the drier foothills surrounding the damp bottomlands of the Willamette Valley. He challenged my assumption and said he had a surprise for me. Richard has provided "artificial cover" for snakes at the refuge by putting out pieces of tin and plywood. I've jokingly told him he's a herpetological litterbug, but he has had spectacular success with this method. By patrolling his litter paths for years, he's amassed invaluable data on snake behavior. Soon, we arrived at the first piece of sun-warmed tin, and upon flipping it over he exposed two Ring-necked Snakes! Astonished, I surveyed our surroundings of a grassy flat with scattered bushes. The site barely rose above a nearby wetland and the closest hill was a good 2 mi (3.2 km) or more away. Additionally, I noted a total lack of rocks, logs, or stumps for natural hiding places. Apparently, the Ringnecks there retreat into rodent burrows and would be difficult to find if not for Richard's introduced retreats. I thanked my friend for jolting me from my preconceived notions and admitted I had learned something new and intriguing that day.

Adult Regal Ring-necked Snake. Near Provo, northwestern, Utah.

Underside of an adult Regal Ring-necked Snake. Near Provo, northwestern Utah.

Smooth Green Snake
Opheodrys vernalis

Adult. Wasatch Range, near Provo, northwestern Utah.

IDENTIFICATION: Adults are 11–26 in (28–66 cm) long. As its common name indicates, this small, slim-bodied snake has smooth dorsal scales and a **dorsal coloration of bright grass green.** The belly is white or pale yellow. **Each nostril is located in the center of a single scale on each side of the head.** Juveniles are gray to olive-gray dorsally.

VARIATION: Two subspecies have been described, with one, the WESTERN SMOOTH GREEN SNAKE (*O. v. blanchardi*), occurring in the Northwest. Some authorities have recently questioned this division and do not recognize either race as valid. It should also be noted that some herpetologists favor placing the Smooth Green Snake in a different genus: *Liochlorophis.*

SIMILAR SPECIES: Some adult Racers (p. 162) may have a greenish or olive tinge to the overall brown dorsal coloration, but it is dull in comparison to the intense green of the Smooth Green Snake. Additionally, the Racer's nostrils are located between scales, rather than in the center of a scale.

DISTRIBUTION: This snake enters our region in the Wasatch and Uinta mountains of northern Utah and some of the higher elevations surrounding the Yampa River drainage of northwestern Colorado and south-central Wyoming. It occurs at elevations from just above 4,000 ft (1,220 m) in the lower foothills and adjoining valley margins, to 7,500 ft (2,290 m) or more in mountainous country. No documented records exist for Utah north of the Salt Lake City–Ogden vicinity, but there are unconfirmed reports for the Logan Canyon area of Cache County. A past record from near Montpelier, Bear Lake County, in southeastern Idaho needs confirmation. Future field work may find that this species also ranges into southwestern Wyoming.

HABITAT AND BEHAVIOR: In western North America, this snake usually inhabits wooded mountains, but it will sometimes enter more open areas along moist riparian zones. In Utah and Colorado, its distribution is closely tied to Gambel Oak—it occurs in grassy glades among scrubby patches of this small tree, often in association with Canyon Maple and Utah Juniper. Although the Smooth Green Snake is diurnal and can sometimes be encountered moving in the open, it is rather secretive and is generally found hiding under rocks, logs, and pieces of bark. It easily vanishes into areas of heavier plant growth near streams and lakes, where its green dorsal color provides excellent camouflage. Smooth Green Snakes also climb and bask on the branches of shrubs and low trees. They eat spiders and adult and larval insects. Little is known about this snake's breeding habits in our region. Elsewhere, it lays 3 to 18 eggs in mid- to advanced summer, sometimes in communal depositories used by several female Smooth Green Snakes. The incubation time for the nearly transparent, thin-shelled eggs varies from one region to the next. In some populations the eggs may be retained within the mother for an extended duration. The embryos are almost completely developed when the eggs are deposited, with hatching taking place in just a few days.

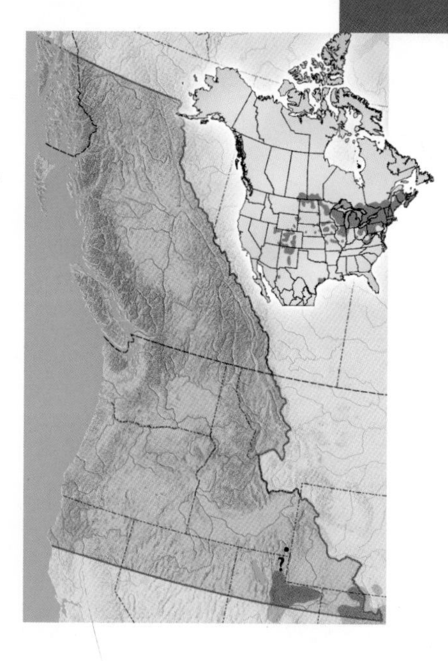

FIELD NOTES: 21 October 1998. Doug Knutsen and I enjoyed a sunny afternoon photographing a Smooth Green Snake among the blazing red maples in Logan Canyon, just east of the Utah State University campus. We had not captured this specimen here, though they are rumored to occur in the area. Dr. Jim MacMahon of the university's science department had recently caught the snake near Provo and generously loaned it to me. Although I much prefer having the direct experience of finding each species of reptile in its particular habitat, I had enthusiastically accepted his offer. Utah herpetologists had warned me that Smooth Green Snakes could be a challenge to locate, and my many failed attempts seemed to confirm their assessment. The closest I had come was finding a dead one on a paved road. In death, the Smooth Green Snake transforms from bright green to dull bluish gray, and it can easily be mistaken for a Racer. No matter how interesting this zoological tidbit was, however, I still needed pictures of a living example. Today, the goal was achieved. Doug and I have helped each other photograph reptiles since we were teenagers, and we work well together for this task. Trying to get good shots of a wriggling snake can be time consuming and exasperating. With Doug's able assistance, though, I had soon exposed three rolls of film. With satisfaction, I checked off another difficult-to-find species from my list.

Adult. Wasatch Range, near Provo, northwestern Utah.

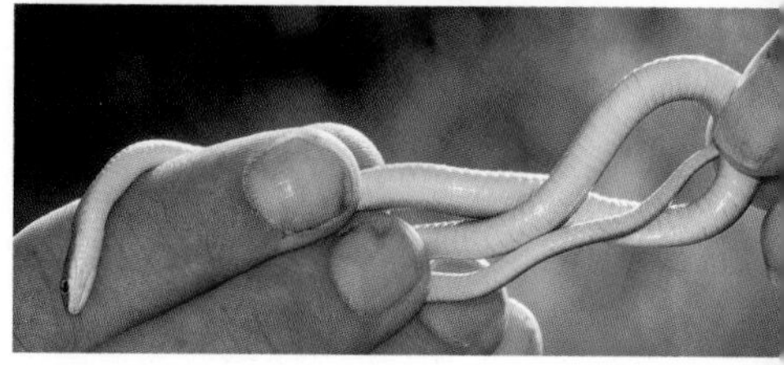

Underside of an adult. Wasatch Range, near Provo, northwestern Utah.

Racer
Coluber constrictor

Adult. Tygh Valley, north-central Oregon.

IDENTIFICATION: Adults are 20–48 in (51–122 cm) long. **The body is slender, with large, smooth scales and a uniform dorsal color of tan, olive, or grayish brown.** Often, there is a bluish or greenish cast to this coloring (particularly along the lower sides of the body), resulting in this snake sometimes being called a "blue racer." **The belly is usually yellow,** although some individuals may have a ventral coloration of cream or white. The throat is white. **The eyes are large and have round pupils.** Juveniles have a row of rust brown blotches along the middle of the back and matching smaller spots on the sides. These markings are most bold at the neck, fading away completely near the middle of the body.

VARIATION: There are 11 subspecies recognized over this snake's coast-to-coast range in the United States and Canada, with the Western Yellow-bellied Racer (*C. c. mormon*) being the race found west of the Rocky Mountains. Racers from northeastern Utah and northwestern Colorado, however, often exhibit characteristics of the Eastern Yellow-bellied Racer (*C. c. flaviventris*). Possibly, the comparatively low gap in the Continental Divide of southwestern Wyoming has allowed some degree of intergradation between these two subspecies in that region.

SIMILAR SPECIES: A blotched juvenile Racer could be confused with a Night Snake (p. 240) or a young Gopher Snake (p. 182). They differ in that the Night Snake has vertical pupils and the Gopher Snake has keeled scales. The Corn Snake (p. 186) also has a similar blotched pattern, but its dorsal scales

are weakly keeled along the back and the blotches continue onto the tail. Additionally, Corn Snakes occur in our region only in northeastern Utah and possibly adjoining northwestern Colorado. A juvenile of the Coachwhip (p. 174) of northwestern Nevada is superficially similar, but it differs in having dorsal crossbars that extend down the sides of the body and often a contrasting neck pattern of black bars on white.

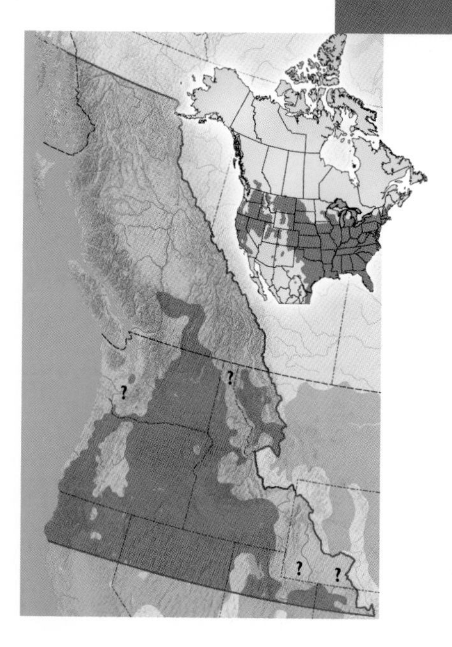

DISTRIBUTION: The Racer is found throughout a good portion of the Northwest at elevations from just above sea level to around 6,000 ft (1,830 m). West of the Cascade Mountains it occurs in both the coastal and inland portions of northern California and southwestern Oregon, but to the north it is confined to Oregon's interior Willamette Valley. Although there are old records for the Puget Sound area of western Washington, that population is now presumed to be extirpated. This snake is widespread in the high plateau country between the Cascades and the Rocky Mountains, ranging from south-central British Columbia to the northern parts of Nevada, Utah and Colorado. Currently, no confirmed records exist for southwestern Wyoming, but it is reportedly found there.

HABITAT AND BEHAVIOR: This reptile occupies a variety of ecosystems, usually preferring exposed sunny situations and avoiding dense coniferous forests. West of the Cascades it inhabits oak woodlands, brushy chaparral, and grassy savannas. To the east of this dividing mountain range, the Racer is found in open juniper and pine forests, rocky canyons and slopes, sagebrush flats, and in the Great Basin Desert. This fast-moving, diurnal predator is often seen hunting with its head elevated, using its excellent eyesight to seek out basking lizards, a primary food source. Small rodents, other snakes, frogs, and birds and their eggs are also eaten. During the summer, its diet often consists primarily of grasshoppers and crickets (especially

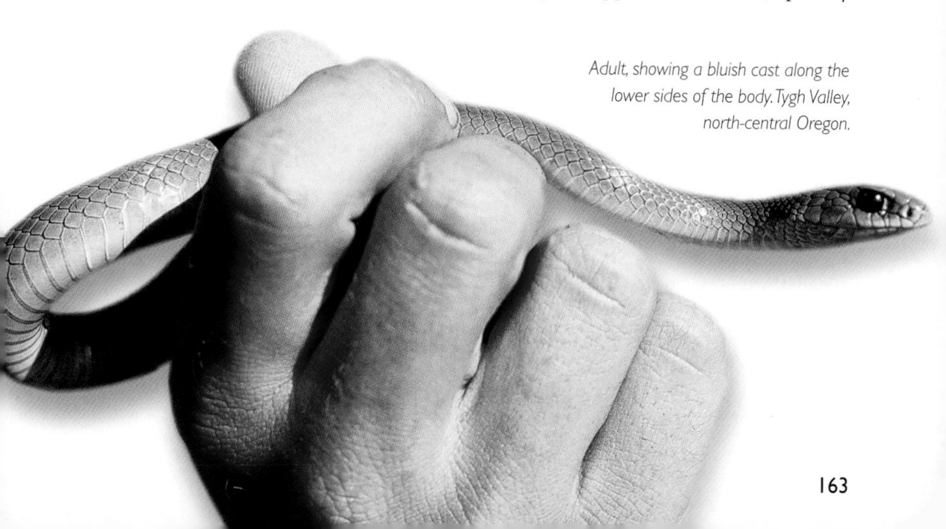

Adult, showing a bluish cast along the lower sides of the body. Tygh Valley, north-central Oregon.

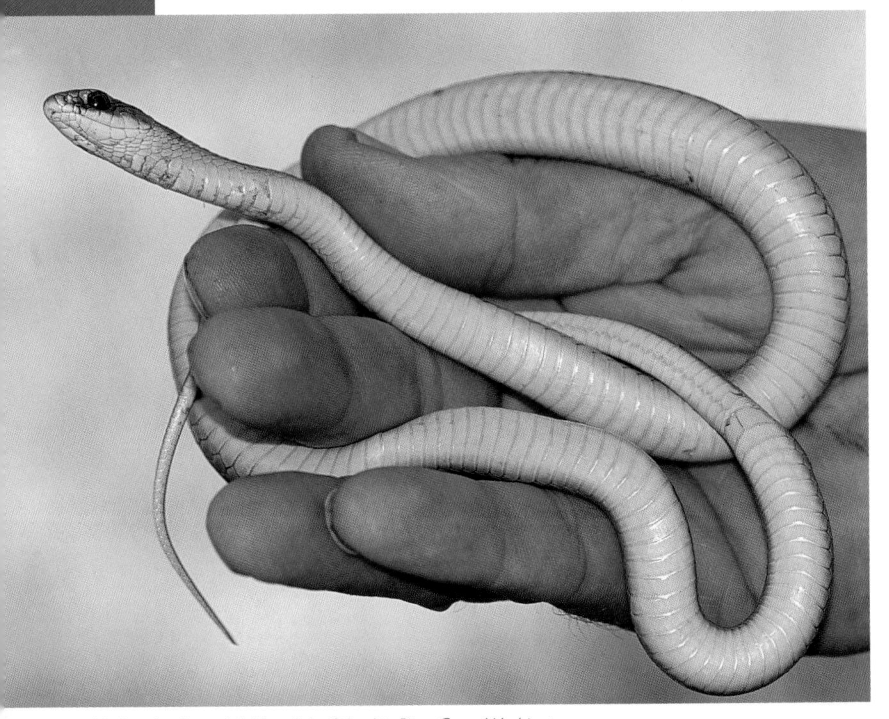

Underside of an adult. Near Lyle, Columbia River Gorge, Washington.

for juvenile Racers). Although usually terrestrial in its habits, it has occasionally been observed climbing in bushes. The name "racer" is appropriate, because this reptile's swiftness makes it difficult to capture. If cornered, it will strike at an enemy and bite fiercely. In the early spring and fall, congregations of Racers are sometimes found at denning sites, often sharing their winter retreats with other species of snakes, including the Western Rattlesnake. From three to seven eggs are laid in late June or early July and hatch during August. Racers are known to sometimes lay their eggs in communal depositories with other species of reptiles at sites that are especially favorable for successful incubation. There is a record from Klickitat County, Washington, of 12 Racer eggs found in a Western Pocket Gopher tunnel.

FIELD NOTES: 9 May 1992. Reptiles were active today during an outing in a rocky canyon on Oregon's side of the Columbia River Gorge. After the cooler weather of early spring, the warm, balmy conditions seemed to be stimulating further dispersal from winter hibernation sites, along with breeding activity and hunting for food. My field companions for the day were my 17-year-old son, Shawn, and biologist Doug Calvin. This section of the spectacularly scenic gorge had lived up to its reputation as an area rich in reptiles. By 2:30 p.m. we had seen several species of lizards and snakes, but the most interesting find of the day was yet to come. Doug had ventured across a small stream to the other side of the canyon and soon summoned us with yells. We hastened through the oaks to where he waited and found him watching something amid grass and spring wildflowers a few feet away. He had discovered a medium-sized Racer that was in the process of swallowing a Western Fence Lizard. The snake was so involved with its meal that it seemed to take little notice of us. I set up my camera and tripod and got some excellent shots of the feeding behavior. Despite this species' scientific name, *Coluber*

Juvenile. Near Hilt, Klamath River drainage, northern California.

constrictor, the Racer does not use constriction to subdue its prey. Instead, the struggling creature being eaten is merely prevented from escaping by a firm grip in the Racer's jaws and sometimes by pinning the prey against the ground under a loop of the body. The snake eventually detected our presence, released the lizard, and disappeared into a clump of poison oak. We felt badly about disrupting its meal, but there were plenty of other lizards basking on the surrounding rocks—the Racer wouldn't stay hungry for long.

Adult eating a Western Fence Lizard. Columbia River Gorge National Scenic Area, near Rowena, Oregon.

Striped Whipsnake
Masticophis taeniatus

Adult; white-striped variation. Alvord Basin, southeastern Oregon.

IDENTIFICATION: Adults are 30–72 in (76–183 cm) long. The body form is extremely slim and sleek, with a long tail, a narrow neck, and an elongated head with large eyes. The basic dorsal color is black, gray, or dark brown with a light stripe of white, cream, or yellow running lengthwise along each side of the back. **A narrow black line (often broken into dashes) runs down the middle of each light stripe. The lower sides of the body are cream or yellowish, with three dividing, black, lengthwise lines.** Ventrally, the belly is cream to pale yellow, changing to coral pink on the underside of the tail. The dorsal scales of the body are large and smooth. **The head scales are distinctively edged with white.**

VARIATION: This snake has an extensive range in western North America, and several subspecies have been described. The DESERT STRIPED WHIPSNAKE (*M. t. taeniatus*) is the only race occurring in the Northwest, and most individuals typify the characteristics assigned to this subspecies. However, there are some differences in dorsal color within two areas of our region that are significant enough to warrant mention. Populations from east of the Wasatch Mountains in northeastern Utah and northwestern Colorado tend to be lighter dorsally, resulting in the dorsal stripes appearing less distinct in contrast to the ground color. This often gives the initial impression of a uniformly pale brown and cream-colored snake that has dashed black lines along the sides of the body. Conversely, those from northwestern California, southwestern and central Oregon and south-central Washington are quite dark with vivid, lemon-yellow stripes.

SIMILAR SPECIES: The California Whipsnake (p. 170) and the Striped Whipsnake are closely related and extremely alike in coloration and pattern. The California Whipsnake differs in not having a narrow black line running down the middle of each light dorsal stripe. The Western Patch-nosed Snake (p. 178) and garter snakes (pp. 210–235) are also striped, but they can be easily distinguished because they all have a middle dorsal

stripe. Additionally, garter snakes are identifiable by their strongly keeled scales.

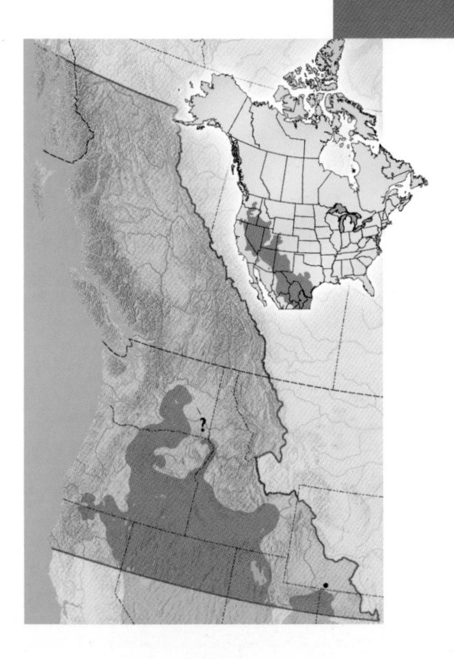

DISTRIBUTION: In the Northwest, the majority of this snake's range is east of the Cascade Mountains, where it is found in northeastern California, northern Nevada and Utah, northwestern Colorado, southern Idaho, central and eastern Oregon, and the Columbia Basin of south-central Washington. There are also populations west of the Cascades in the interior sections of Oregon's Rogue River drainage and the Klamath River–Shasta Valley areas of northwestern California. There is one sight record for southwestern Wyoming in Sweetwater County.

HABITAT AND BEHAVIOR: The Striped Whipsnake inhabits brushy country at elevations from just above sea level along the Columbia River to near 7,000 ft (2,130 m) in northern Utah and Colorado. Although

Adult; yellow-striped variation. Near Hilt, Klamath River drainage, northern California.

Adult; pale variation. Near Duchesne, Uinta Basin, northeastern Utah.

Adult climbing in a shrub. Near Duchesne, Uinta Basin, northeastern Utah

Underside of an adult, showing coral pink color on the tail. Alvord Basin, southeastern Oregon.

usually encountered in rocky areas of slopes and canyons, it is also sometimes seen on open flats. In the interior plateau region, the habitats occupied include Great Basin saltbush-greasewood communities, juniper-sagebrush associations, and, at the southeastern perimeter of its range, pinyon pine–juniper forests. In the Klamath Mountain region of southwestern Oregon and northwestern California this reptile is found in mixed woodlands of Oregon White Oak, California Black Oak, Pacific Madrone, and Ponderosa Pine with a shrub understory of White Manzanita, Buckbrush, Deerbrush, and Bitterbrush. Exceedingly swift-moving and an excellent climber, the Striped Whipsnake often escapes by gliding through the tops of bushes. If cornered, it will assume a threatening defensive posture with the head and fore portions of the body protruding from the branches, and it will aggressively bite to protect itself. The combined speed and keen eyesight of this species make it a formidable diurnal predator. Lizards form the major portion of the diet, but small mammals, snakes (including rattlesnakes), nestling birds, frogs, and insects are also eaten. A northern Utah study found that winter denning sites are often shared with Racers, Gopher Snakes, and Western Rattlesnakes. Upon emergence from hibernation in the spring, males actively seek out females for mating and will engage in entwining male combat struggles. From 3 to 10 eggs are laid in late June or early July (sometimes communally with other snake species) and hatch in late August or early September.

FIELD NOTES: 26 September 1997. Capturing snakes can sometimes be painful to both the body and ego, as today's incident near the Alvord Desert in southeastern Oregon proved. Accompanied by my family and our friends, Laura McMasters and her grandson Sebastian, we took an evening drive from camp to the tiny community of Fields for gas and ice. Because it was near closing time for the general store there, we didn't have time to stop for a large Striped Whipsnake that was basking in the sunshine on the gravel road. On the return trip, however, we were delighted to see that the reptile hadn't

Juvenile. Lower Metolius River Canyon, central Oregon.

moved. Just as we were cautiously approaching the wary snake, a young man and woman also pulled over in a pickup and stopped. As it turned out, they worked for the Oregon Museum of Science and Industry and were taking an interest in the proceedings. Wanting to give everyone a close-up look at this species, I decided that I would demonstrate to my impromptu audience how an experienced herpetologist could adeptly catch the elusive Striped Whipsnake. I knew that the larger and more threatening I appeared, the greater the chance of spooking the turbo-charged creature. Therefore, I got down on my hands and knees and had soon managed to get within arm's length of it. The snake had raised its head several inches off the road and was intently watching me as I inched my hand outward. In short order, though, it decided to put an end to this foolishness by turning to speed away. I lunged forward onto my belly and successfully grabbed the snake's tail, which precipitated an immediate response. The next thing I saw was a wide-open snake mouth full of teeth coming toward my right eye, and before I knew what hit me was bitten on the cheek. While blood ran down the side of my face into my beard, everyone admired the lovely markings and coloration of the wriggling snake in my hands.

This Striped Whipsnake was in a characteristic pose, with its head elevated, as it alertly watched my stealthy approach. Alvord Basin, southeastern Oregon.

California Whipsnake
Masticophis lateralis

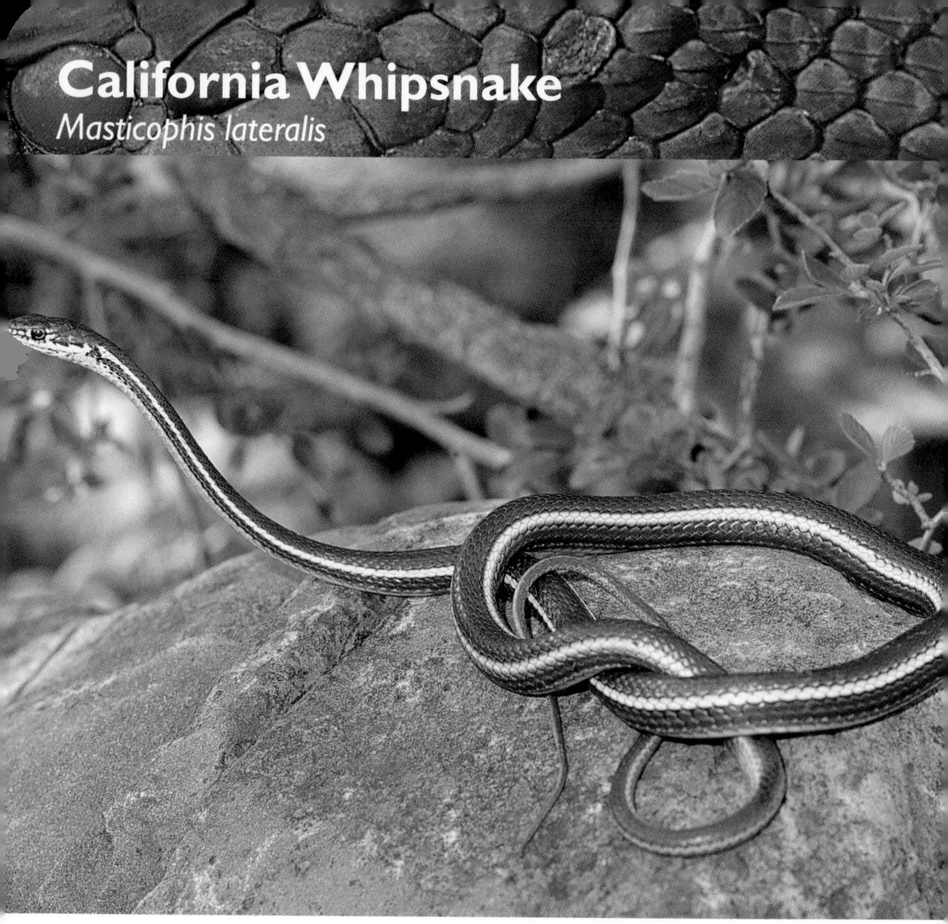

Adult. The Nature Conservancy's Dye Creek Preserve, near Red Bluff, Sacramento Valley, northern California.

OTHER NAMES: Striped Racer.

IDENTIFICATION: Adults are 30–60 in (76–152 cm) long. This slender, streamlined snake has **a dorsal ground color of dark brown to black, with a conspicuous light yellow to cream stripe running lengthwise along each side of the back. Both light stripes are solid, not bisected down the middle by a narrow black line.** The belly is pale yellow to cream in color, and the underside of the tail is coral pink. The head is narrow and elongated, with noticeably large eyes. The dorsal scales are large and smooth.

VARIATION: Two subspecies are recognized, with one, the CHAPARRAL WHIPSNAKE (*M. l. lateralis*), occurring in our area.

SIMILAR SPECIES: The closely related Striped Whipsnake (p. 166) has a nearly identical pattern and coloration, but can be differentiated by a narrow black line that runs along the middle of each light dorsal stripe. Garter snakes (pp. 210–235) also have a dorsal pattern of light, lengthwise stripes, but they differ in having keeled scales and usually a third stripe down the middle of the back.

DISTRIBUTION: Found throughout much of California, this snake enters the area covered by this guide in the Sacramento River drainage. Locality records indicate that it reaches the northern limits of its range in extreme southern Siskiyou County just south of Dunsmuir, and along the Trinity River at Clair Engle Lake, Trinity County. In our region, it occurs at elevations from just over 400 ft (120 m) near Red Bluff in the Sacramento Valley to 3,000 ft (910 m) in the vicinity of Whitmore on the western slope of the Sierra Nevada.

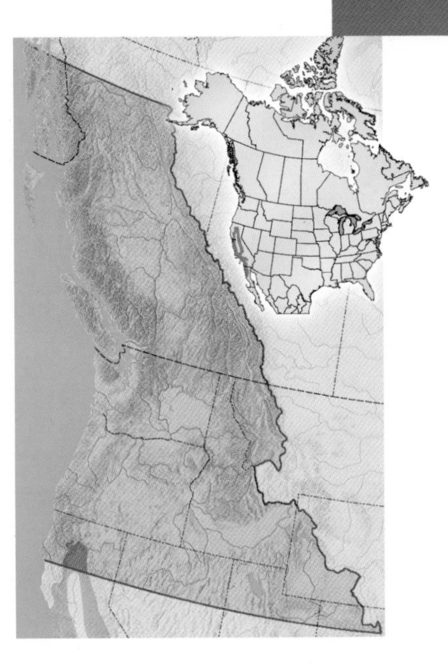

HABITAT AND BEHAVIOR: The California Whipsnake primarily inhabits brushy chaparral of the rocky foothills, often being seen along streamcourses. It will sometimes range out into the Blue Oak savannas of the valley floor where the appropriate chaparral habitat of Buckbrush and manzanita species grows along ravines in riparian areas. It is also found in the more open, brushy sections of surrounding mountains in mixed woodlands of Oregon White Oak, California Black Oak, Canyon Live Oak, Pacific Madrone, Gray Pine, and Ponderosa Pine. Like all species of whipsnakes in the *Masticophis* genus, this diurnal reptile is fast, agile, and high-strung, commonly holding its head in an elevated position when on the alert. Anyone trying to capture a California Whipsnake will usually experience a lively chase as it adeptly moves through nearly impenetrable thickets of brush, even ascending trees in an effort to escape. Its climbing ability is frequently used for searching the branches of shrubs and trees for prey, and it is occasionally encountered as it basks on a sunny limb. Lizards are the major food source, but snakes (including rattlesnakes), small mammals, frogs, birds and their nestlings, and insects are additional components of the diet. Little is known concerning the breeding habits of this species in northern California. Records from throughout the state show that from 6 to 11 eggs are laid. At our latitude, eggs are probably deposited in June or July and hatch during August or early September.

Adult. The Nature Conservancy's Dye Creek Preserve, near Red Bluff, Sacramento Valley, northern California.

California Whipsnake

Underside of an adult, showing coral pink color on the tail. The Nature Conservancy's Dye Creek Preserve, near Red Bluff, Sacramento Valley, northern California.

FIELD NOTES: 15 June 1998. The goal of my hike today at The Nature Conservancy's Dye Creek Preserve in Tehama County, California, was to find and photograph a California Whipsnake in its habitat. It is always a fascinating challenge to delve into the available literature, learn about the natural history of a species I'm unfamiliar with, and then go out and try to find it. Despite my preparations, I'd had repeated failures throughout northern California owing to a cool, wet spring season. I was optimistic, though, that this sunny afternoon would be different. Marvin Abst, of the Biology Department at Shasta College in Redding, had advised me that the brushy borders along streams were good places to find this snake. I zeroed in on the creek, and within half an hour I had progressed a considerable distance up the canyon. I had just moved out of the shade of some California Sycamores, into an area where the sun glinted off the flowing water,

when I noticed a movement on the other side of the stream. It was a large California Whipsnake slithering toward a thick clump of Buckbrush! I nearly flew across Dye Creek, hopping from rock to rock, but by the time I got there the snake had disappeared into the shrubs. Initially, I thought my chance was missed, but then a fleeting memory came to mind. I had recently read a 1949 account by Henry S. Fitch in which he made an observation about this species: "Sometimes, at an alarm, one would dart through a bush to its farther side and lie motionless there until closely approached." Thinking this worth a try, I carefully crept around the screening Buckbrush and peered behind it. Sure enough, there was the snake. A quick grab and I had the beautifully striped serpent in my hands. Here was a good example of the usefulness of keeping accurate field notes on wildlife behavior, even when the observations are half a century old. Thank you Dr. Fitch.

Adult climbing in a shrub. The Nature Conservancy's Dye Creek Preserve, near Red Bluff, Sacramento Valley, northern California.

Coachwhip
Masticophis flagellum

Adult. Pyramid Lake, northwestern Nevada.

IDENTIFICATION: A good-sized, lithe snake, adults are 36–72 in (91–183 cm) long or more. The dorsal scales are large and smooth. The general body form is slim, with a narrow head that has big, prominent eyes. The common name is derived from a **pattern of dark edging on the scales of the rear portions of the body and the long, slender tail, which creates a resemblance to a braided leather whip. There are dorsal markings consisting of brown to reddish-brown, irregular crossbars that are brightest on the neck (sometimes nearly black) and progressively become more faint toward the mid-body. The rear half or third of the snake is usually a uniform light brown, tan, or yellowish tan.** The belly coloration may be cream, yellowish, pinkish, or pale brown, with a scattering of darker spots and blotches. Juveniles have a pattern of dark crossbars along the back that are more distinct than those of adults and usually extend the entire length of the body. Additionally, there is often a contrasting pattern on the neck of black bars on white.

VARIATION: There are considerable differences in coloration and pattern across this snake's coast-to-coast distribution in North America and several subspecies are recognized. Only one, the Red Coachwhip (*M. f. piceus*), occurs in our region. In the American Southwest, this species' dorsal coloration is often intensely red or pinkish, and it is sometimes called the "red racer." In the northern part of its range, however, the Coachwhip is basically brown or tan, with only a hint of red in the dark brown crossbars on the neck.

SIMILAR SPECIES: No other species of snake native to the Northwest has brown crossbars on the back that fade away toward the rear of the body.

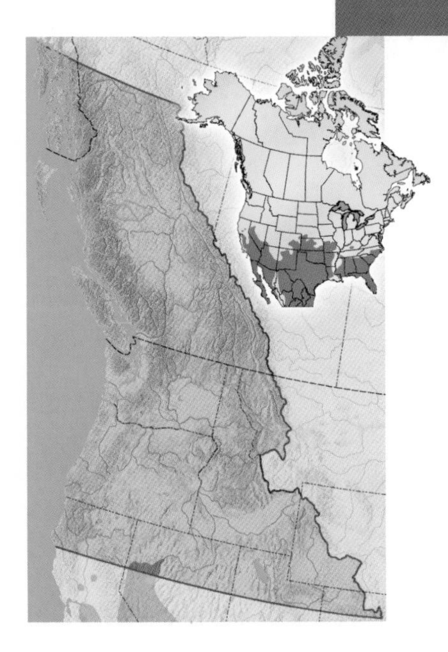

DISTRIBUTION: The Coachwhip reaches the northern limits of its range in the desert basins of northwestern Nevada, at elevations of 3,500–4,500 ft (1,070–1,370 m). Within the region covered by this book, it has been recorded only in the vicinity of Pyramid Lake, the Winnemucca Lake Basin, and the Lovelock area as far north as Rye Patch Reservoir. It may range farther north into the Smoke Creek Desert and Black Rock Desert, as well, but there are no confirming records at present. It is probable that future field surveys will find that this snake also occurs in the Honey Lake Basin of adjacent California.

HABITAT AND BEHAVIOR: In our region, the Coachwhip is restricted to the arid Great Basin Desert ecosystem in sunny, exposed areas with a covering of brush. Typical shrubs associated with its habitat are Big Greasewood, Bailey's Greasewood, Shadscale, Four-winged Saltbush, and Green Ephedra. It seems equally at home on both rocky slopes and the sandy flats of basins. This diurnal predator has a tolerance for high temperatures, and it is often seen abroad at midday during the heat of summer. It is extremely active and is usually encountered as it moves through brush and rocks hunting prey or as it crosses a road. The head is frequently raised high off the ground, with the large eyes alertly surveying the surroundings. The Coachwhip is fast moving, and when pursued it seems to virtually disappear before

Adult. Pyramid Lake, northwestern Nevada.

Underside of an adult. Pyramid Lake, northwestern Nevada.

one's eyes in a matter of seconds as it effortlessly glides away into a hiding place (rodent burrows are frequently used). It is also an excellent climber and will often escape into the branches of large bushes or, if present, trees. When cornered, this snake readily defends itself by striking and biting, using a slashing style with its teeth that is effective in tearing the skin of a would-be captor. If sufficiently provoked, it will sometimes aggressively advance toward a perceived enemy. The Coachwhip feeds on small mammals, lizards, snakes, birds and their eggs, and large insects (grasshoppers, crickets, and cicadas). The reproductive habits for this species in our region are unknown. Elsewhere within its range in the United States, it is reported to lay 4 to 20 eggs. At the latitude of northwestern Nevada, eggs are probably deposited in July and hatch in late August or early September. The author observed a recently hatched Coachwhip at Pyramid Lake on 11 September 1999.

FIELD NOTES: 2 June 2000. Herpetologists usually agree that the most difficult to capture snake in North America is the Coachwhip. The speed and wariness of this species is legendary. This was certainly my experience during four years of trying to obtain an adult specimen in northwestern Nevada to photograph for this book. On

Juvenile. Pyramid Lake, northwestern Nevada.

eight different occasions, Coachwhips had mercilessly humbled me by escaping with seemingly utter ease. One episode was especially ego-deflating. For half an hour I pursued a 5-ft (1.5-m) individual that I discovered stalking the same Desert Spiny Lizard I was photographing. The Coachwhip adeptly eluded my grab for its tail as it chased the lizard around a rock outcrop and then ducked into a kangaroo rat burrow. I spent several minutes excavating the rodent tunnel, but the snake suddenly popped out another entrance and vanished before I could do anything sensible. The only hiding place it could have quickly used was a nearby boulder. I heaved upward on the large chunk of stone, painfully wrenching my right shoulder, and the Coachwhip shot out and sped away with me in hot pursuit. I made a desperately unsuccessful lunge and ended up flat on my belly with my face in the sand. I felt like a participant in a *Roadrunner and Coyote* cartoon! Today, when I encountered a Coachwhip and it hid in a kangaroo rat burrow, I profited by the lesson learned from my past mistake. Four holes led into the shallow tunnels, so I plugged three with rocks and began digging into the fourth opening. I soon unearthed the snake's tail and grabbed it. Swift as lightning, the other end of the snake came out and bit me on the hand and I grasped it by the neck. The following morning, after chilling the snake to a lethargic state in my ice chest, I exposed several rolls of film and achieved my goal at last. Afterwards, with genuine respect for this graceful reptilian predator, I watched it slither away to freedom.

Western Patch-nosed Snake

Salvadora hexalepis

Adult. Pyramid Lake, northwestern Nevada.

IDENTIFICATION: Adults are 20–46 in (51–117 cm) long. **There is a distinctively large, triangular-shaped rostral scale that curves back over the top of the snout.** Dorsally, the basic ground color is light grayish tan. A broad (3 scale rows wide) lengthwise beige or yellowish stripe runs along the middle of the back and is bordered on each side by a row of dark brown to nearly black crossbars. The crossbars often unite to form a dark, lengthwise band, particularly at the rear of the body and on the tail. Another narrower, dark, lengthwise band (often discontinuous or faint) usually runs along the lower sides of the body. The dorsal scales are large and smooth. The belly is white, sometimes shading to pale orange toward the tail. Males have keeled scales just above the vent and ventrally at the base of the tail.

VARIATION: Four subspecies are recognized, with only one occurring in our area: the DESERT PATCH-NOSED SNAKE (*S. h. hexalepis*).

SIMILAR SPECIES: The Striped Whipsnake (p. 166) coexists in the same habitat of northwestern Nevada and the adjacent Honey Lake Basin of northeastern California, but it differs in not having a dorsal stripe down the middle of the back. Garter snakes (pp. 210–235) usually have a middle dorsal stripe, but they are easily identified by their strongly keeled dorsal scales.

DISTRIBUTION: The northern limit of the Western Patch-nosed Snake's range is in northwestern Nevada at elevations of 3,500–4,500 ft (1,070–1,370 m). Within the region covered by this guide, it has been

recorded from Pyramid Lake, the Winnemucca Lake Basin, and the Smoke Creek Desert, and in the Black Rock Desert as far north as the area of Sulphur. It may also occur in the Lovelock vicinity, but there are no confirming records at present. Additionally, this snake has been found in adjacent northeastern California in the Honey Lake Basin of Lassen County.

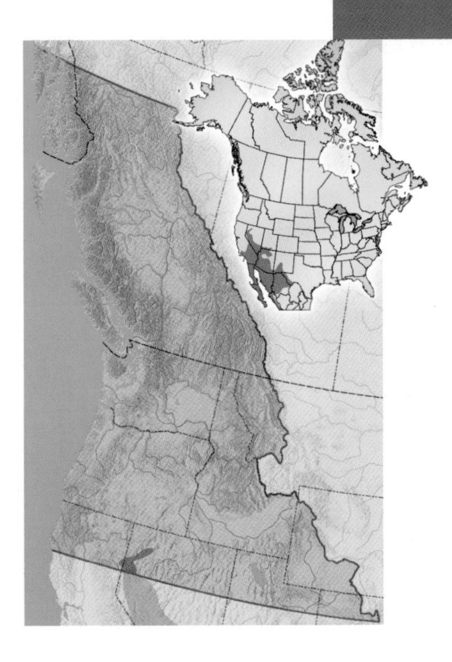

HABITAT AND BEHAVIOR: The Western Patch-nosed Snake inhabits the arid Great Basin Desert ecosystem in sunny, exposed areas that have a covering of brush. Shrubs associated with its habitat are Big Greasewood, Bailey's Greasewood, Shadscale, Four-winged Saltbush, Green Ephedra, and Big Sagebrush. It is a diurnal reptile that is seen in both the rocky areas of slopes, washes, and canyons, as well as in sandy basin flats. Like its close relatives, the Racer and the whipsnakes, this is a wary, fast-moving snake that adeptly escapes into brush when chased. Although not a true burrower like the Long-nosed Snake or the Ground Snake, it seems to use the enlarged rostral scale on the snout when digging for lizard and snake eggs, its favored foods. Lizards and small mammals form the major portion of its diet. The breeding habits of this species are unknown for our region. Elsewhere in its range, the meager data concerning its reproduction indicate that 4 to 10 eggs are laid during late

Adult. Pyramid Lake, northwestern Nevada.

Underside of an adult. Pyramid Lake, northwestern Nevada.

spring or early summer and hatch in late summer or early fall.

FIELD NOTES: 3 June 2000. Knowing the concrete facts about a reptile's habitat preferences and behavior is indispensable when trying to locate and observe a particular species. Nevertheless, at the risk of sounding a bit mystical, I have to admit that intuition sometimes plays an important role, as well. Other than finding a few dead specimens on roads, I had never seen a live Western Patch-nosed Snake in the wild. Because I had encountered most of the road kills in the area of Nevada's Pyramid Lake, I kept returning over a four-year period. In the warm mornings and late afternoons of spring and summer, I would leave camp and drive the various surrounding roads in the hope of finding a crossing individual of this species. Despite logging miles and miles of roadhunting, none turned up. During my many slow drives, I kept noticing one particular location along a dirt road above the shore of the lake where a promising looking jumble of rocks and brush approached the roadside. For some unexplainable reason, I had a "gut-feeling" that I'd eventually find a Western Patch-nosed Snake there, so I kept returning to the site every time I visited Pyramid Lake. Today, my perseverance bore out this hunch. The morning dawned under a crystal-clear blue sky. With an uncanny sense of complete confidence that I'd be successful, I immediately patrolled the section of road by my intuitionally tagged spot. Approaching the area, I could see a snake stretched across the left road rut. There was my Western Patch-nosed Snake waiting for me! With gratification, I got my needed photographs of it and added a new species to my reptile life list.

Juvenile. Black Rock Desert, northwestern Nevada.

Gopher Snake

Pituophis catenifer

Adult Pacific Gopher Snake. The Nature Conservancy's Dye Creek Preserve, near Red Bluff, Sacramento Valley, northern California.

OTHER NAMES: Bullsnake.

IDENTIFICATION: A large, heavy-bodied snake, adults range from 30 in (76 cm) to slightly over 72 in (1.8 m) in length. There is a row of dark brown, reddish-brown, or black blotches along the middle of the back, with corresponding smaller blotches along the sides of the body. The tail is marked with dark, encircling rings. The overall ground color may be tan, brown, yellowish, cream, or gray, occasionally with a reddish-orange tinge. A dark facial mask stripe crosses the top of the head and passes through both eyes. **The pupils of the eyes are round.** Because the rostral scale is somewhat enlarged, the snout is slightly cone-shaped in appearance. **Dorsally, the scales of the back are strongly keeled, but**

they grade into smooth scales on the lower sides of the body. The belly is white or pale yellow, usually with scattered black spots.

VARIATION: Several subspecies are recognized across this snake's wide range in North America, with three being found in our region. The PACIFIC GOPHER SNAKE (*P. c. catenifer*) has a grayish-brown suffusion on the sides of the body, and the dark dorsal blotches on the neck are usually well separated and not connected with the smaller blotches on the lower sides. It occurs west of the Cascade Mountains. The GREAT BASIN GOPHER SNAKE (*P. c. deserticola*) has light tan or cream sides without the grayish-brown suffusion. The dark dorsal blotches on the neck are usually connected at the lower sides by smaller, elongated blotches or stripes, causing the light interspaces between the dorsal blotches to be surrounded by dark markings. It occurs east of the Cascade Mountains, with the exception of western Montana. There is a broad area of intergradation between the Pacific Gopher Snake and the Great Basin Gopher Snake along the eastern slope of the Cascades in central Oregon and northeastern California. Expect to find individuals with characteristics of both forms in that region. The BULLSNAKE (*P. c. sayi*) has a high, narrow rostral scale, giving the snout a pointed appearance. The dark dorsal blotches on the neck are usually well separated and not connected with the smaller blotches on the lower sides. The body form is quite stout, often with a yellowish or reddish-orange ground color and without a grayish-brown suffusion on the sides of the body, and the tail is relatively thick, with broad, dark, encircling rings. It occurs in western Montana.

SIMILAR SPECIES: There are four species of Northwestern snakes with similar blotched patterns. They can be differentiated from the Gopher Snake by the following characteristics: the Night Snake (p. 240) and the Western Rattlesnake (p. 244) both have vertical pupils; the juvenile Racer (p. 162) has smooth scales on the back; and the Corn Snake (p. 186) has weakly keeled scales on the back, a distinctive V-shaped marking on top

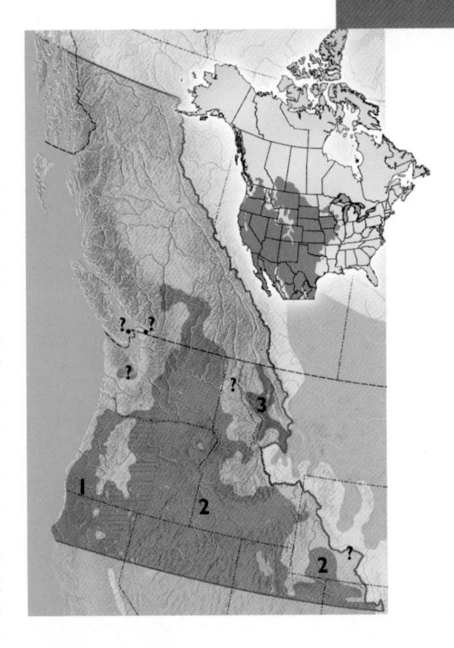

SUBSPECIES

- 1 — Pacific Gopher Snake *P. c. catenifer*
- Area of intergradation
- 2 — Great Basin Gopher Snake *P. c. deserticola*
- 3 — Bullsnake *P. c. sayi*
- Species range outside the Northwest

Adult Pacific Gopher Snake; darker Willamette Valley variation. Wilson Wildlife Area, near Corvallis, northwestern Oregon.

Gopher Snake

Adult Great Basin Gopher Snake. Black Rock Desert, northwestern Nevada.

Adult Great Basin Gopher Snake, showing the interconnected dark neck markings typical for this subspecies. Owyhee River Canyon, southeastern Oregon.

Adult Bullsnake. National Bison Range, western Montana.

of the head, and occurs in our area only in the Green River drainage of northeastern Utah.

DISTRIBUTION: The Gopher Snake is found throughout most of the Northwest at elevations of just above sea level to nearly 6,000 ft (1,830 m), exclusive of moist, densely forested mountain ranges and coastal areas. West of the Cascades it is common in the Klamath Mountains region of northern California and southwestern Oregon, along with the Willamette Valley of northwestern Oregon. Records from the 1800s for this species in the Puget Lowlands of western Washington and southwestern British Columbia have never been duplicated and may be in error or indicate that the Gopher Snake is now extirpated there. East of the Cascade Mountains, this reptile occurs nearly everywhere in the interior plateau country to the Rocky Mountains, and from south-central British Columbia southward through Nevada and Utah.

HABITAT AND BEHAVIOR: Preferring more open, dry terrain, the Gopher Snake inhabits oak savannas, brushy chaparral, meadows, and the sparse, sunny sections of coniferous forests in the western portions of California and Oregon. In the arid and semiarid inland plateaus it ranges from deserts through sagebrush steppe and grassy prairies, to juniper and pine woodlands. This snake is also commonly encountered at the margins of farmlands and around barns and outbuildings. As the common name suggests, it is agriculturally valuable, eating gophers and other small mammals, such as mice, rats, moles, squirrels, and small rabbits. Lizards and the eggs and nestlings of birds are also occasionally eaten. Constriction is used to subdue prey. Sometimes known as the "blowsnake," the Gopher Snake will hiss loudly when it feels threatened. Additionally, it will often flatten its head, coil and strike defensively, and nervously vibrate its tail. If the snake happens to be among dry leaves, the tail may vibrate against these and produce a faint whirring sound. This behavior frequently convinces people that it is a venomous rattlesnake and many harmless Gopher Snakes are needlessly killed. During the spring and autumn this snake is largely diurnal, but it becomes increasingly

Juvenile Pacific Gopher Snake. Wilson Wildlife Area, near Corvallis, Willamette Valley, northwestern Oregon.

crepuscular and nocturnal when the hot days of summer arrive. At that time, many Gopher Snakes are frequently seen crossing roads in the evenings and at night. From 4 to 20 eggs are laid in rodent burrows or rock crevices in late June or early July. Hatching takes place in late August or early September.

FIELD NOTES: 29 July 1995. Craig Zuger and I observed a good example today of how individual snakes of the same species can sometimes differ greatly in temperament. We had taken an afternoon hike in the Trout Creek Mountains of southeastern Oregon, and evening shadows were lengthening by the time we had driven a jeep road back to the paved highway. When we paused at the stop sign, I happened to glance out the car window and noticed two medium-sized Gopher Snakes sunning next to a bush at the roadside. We got out and approached them, triggering two entirely different reactions in the reptiles. One snake merely raised its head slightly and peered in our direction when it noticed us, while the other immediately threw itself into a tightly coiled defensive pose and began hissing loudly. Craig picked up the mellow snake and it allowed itself to be held with no struggle whatsoever. Conversely, when I reached toward the other Gopher Snake it began violently striking at me. As soon as I picked it up, however, the snake immediately quit trying to bite and became relatively relaxed in my grasp. Most Gopher Snakes are like this: high-strung at first but usually acquiescing when handled. The occasional individual, though, seems totally unconcerned with being captured.

Underside of an intergrade between the Pacific and Great Basin subspecies. Columbia River Gorge National Scenic Area, near Rowena, Oregon.

A cranky Gopher Snake flattening its head, coiling, hissing, vibrating its tail, and striking defensively. Alvord Basin, southeastern Oregon.

Corn Snake

Elaphe guttata

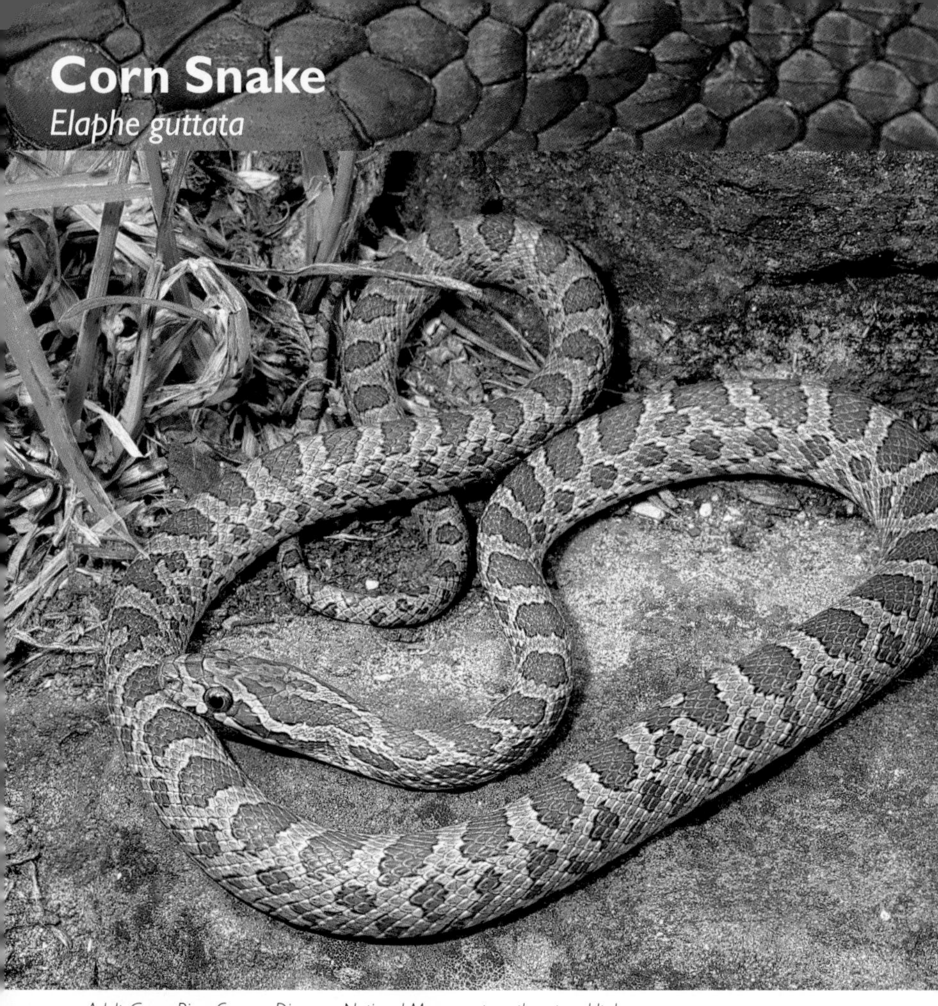

Adult. Green River Canyon, Dinosaur National Monument, northeastern Utah.

IDENTIFICATION: In our area, adults are 18–36 in (45–91 cm) long. The dorsal coloring and pattern consist of a light gray background with brown to olive-brown, narrow, oval blotches along the back that are edged in black. A series of smaller brown blotches run along the sides of the body. There is a brown facial mask stripe across the head that extends through each eye. **Two larger brown stripes begin on the neck and more or less converge on the top of the head to form a distinctive V.** The pupils of the eyes are round. Most of the dorsal scales are smooth, but those on the back are weakly keeled. The belly is white, with many squarish black blotches, and flat, creating an abrupt, angular edge at the sides of the body. The underside of the tail is often marked with black stripes.

VARIATION: There are three subspecies, with one occurring in the Northwest: the GREAT PLAINS RAT SNAKE (*E. g. emoryi*). West of the Continental Divide, individuals of this subspecies are smaller and tend to be paler gray with less vivid dorsal markings. Some herpetologists consider these differences to be significant enough to warrant designating these populations as a separate subspecies, the INTERMOUNTAIN RAT SNAKE (*E. g. intermontana*).

SIMILAR SPECIES: Four other species of blotched snakes coexist with the Corn Snake in our area, but they can be differentiated by the following characteristics: the Gopher Snake (p. 182) has strongly keeled scales on the back; the juvenile Racer (p. 162) has smooth

scales on the back; and the Night Snake (p. 240) and Western Rattlesnake (p. 244) have vertical pupils. In addition, all these species lack the Corn Snake's distinctive, V-shaped marking on the top of the head.

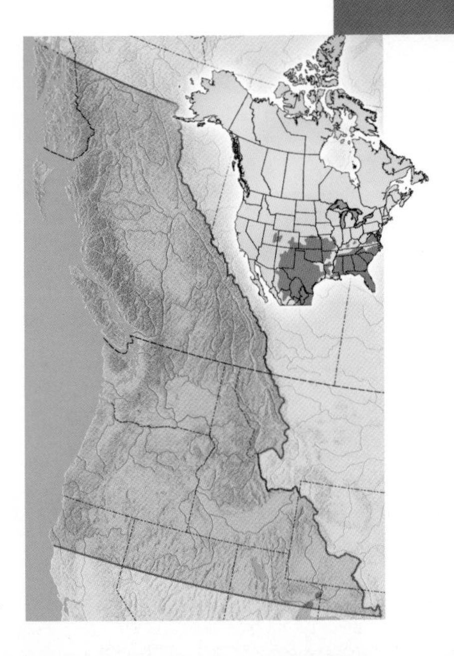

DISTRIBUTION: Although the primary range of this snake is in the southern Great Plains and eastern portions of North America, disjunct populations occur west of the Rocky Mountains in the Colorado River drainage of western Colorado and eastern Utah. The Corn Snake enters the area covered by this guide along the Green River at elevations of 4,500–6,000 ft (1,370–1,830 m). It has been recorded at only two localities in Uintah County, Utah. One site is a few miles south of Ouray and the other is at Dinosaur National Monument. It is probable that future field work will find this species at other locations in northeastern Utah and possibly along the Green and Yampa rivers of adjacent northwestern Colorado.

Adult, showing the distinctive V-shaped markings on the top of the head. Green River Canyon, Dinosaur National Monument, northeastern Utah.

Underside of an adult. Green River Canyon, Dinosaur National Monument, northeastern Utah.

HABITAT AND BEHAVIOR: In the arid West, the Corn Snake is largely restricted to riparian areas in rocky canyons and valley bottoms where there is a growth of willows, cottonwoods, junipers, sagebrush, and rabbitbrush. It is also sometimes found at the edges of irrigated farmland. The common name might be derived from markings on the belly that resemble the checkered pattern of Indian corn. During the warm part of the year, the Corn Snake becomes active during the evening and at night, when it is occasionally observed crossing roads. Otherwise, it is seldom seen and remains hidden in rodent burrows and under surface objects during the day. This snake is an excellent climber, but it is generally encountered on the ground. When threatened, it usually holds the forward portion of its body in an elevated defensive coil, and it will strike and attempt to bite. If captured and handled, it discharges a foul-smelling musk and fecal matter. It eats small mammals, birds, lizards, and frogs, subduing them by constriction. Little is known about the Corn Snake's reproductive habits west of the Continental Divide. A female specimen collected during early July in southeastern Colorado contained nine eggs that appeared to be ready for laying.

FIELD NOTES: 17 July 1998. All days in the field should be as successful as this one. With 45 species of reptiles to find and photograph for this book project, I had good reason to worry that some especially rare varieties would completely elude me. The Corn Snake was at the top of my list of challenges. Only two of these snakes had ever been found within the boundaries of the region covered by this guide. One of the localities was in Dinosaur National Monument in northeastern Utah. It is a ruggedly scenic area and I had been looking forward to hiking into the remote section of the

Green River Canyon where the snake had been collected in 1982. By mid-afternoon the air temperature had climbed to nearly 100° F (38° C), but I had timed my excursion so I would arrive at my destination during the cooler evening, when Corn Snakes would begin to prowl. At the trailhead I topped off my water bottle, shouldered my camera pack, and optimistically struck out along the 4-mi (6.4-km) path that winds down a side canyon to the Green River. I was soon following a rushing stream in the evening shadows below craggy pink cliffs that glowed in the fading sun. Within an hour and a half I was nearing the river in the main canyon, when I happened to look downward just in front of my feet. There, barely discernible in the dim light of dusk, was a snake crossing the path. I reached down, picked it up, and joyfully gave a triumphant shout when I saw that it was a Corn Snake. Using a flash, I took a number of photographs of the rare reptile, released it, and hiked back out of the canyon by flashlight. This excursion had almost seemed too easy and was a welcome change from the more typical days of fruitless searching for often elusive reptiles.

When the summer sun sets below the horizon, Corn Snakes begin to prowl. Twilight in the Green River Canyon of Dinosaur National Monument, Utah.

Common Kingsnake

Lampropeltis getula

Adult. Near Hilt, Klamath River drainage, northern California.

IDENTIFICATION: Adults are usually 24–45 in (61–114 cm) long in our region. This strikingly marked reptile is easily identified by its **bold, alternating pattern of black and white crossbands that completely encircle the body.** The white bands widen at the sides of the body. These markings are more pale and irregular on the belly. The top of the head is entirely black, except for one small spot of white at the rear, while the nasal and lip scales are white. The dorsal scales are smooth and very glossy. Some individuals in the Sacramento River drainage of northern California have dark chocolate brown bands, instead of black, and occasionally the light bands are cream or yellow instead of white.

VARIATION: Seven or eight subspecies are recognized across this snake's coast-to-coast range in the United States. Only one is found in the Northwest: the CALIFORNIA KINGSNAKE (*L. g. californiae*).

SIMILAR SPECIES: The Ground Snake (p. 236) and the Long-nosed Snake (p. 206) both have variations with black-and-white dorsal crossbanding. The ranges of these two species overlap with that of the Common Kingsnake in the Great Basin of northwestern Nevada and the adjoining Honey Lake Basin of northeastern California. These two snakes can be differentiated from the Common Kingsnake by the fact that their bands do not cross the belly.

DISTRIBUTION: The Common Kingsnake reaches the northern limits of its distribution in western North America in our region. West of the Cascade Mountains, it ranges from northern California into the Rogue and Umpqua river drainages of southwestern Oregon, at least as far north as Elkton in Douglas County. To the east, this snake occurs in the Great Basin of Nevada and northeastern California (Honey Lake Basin).

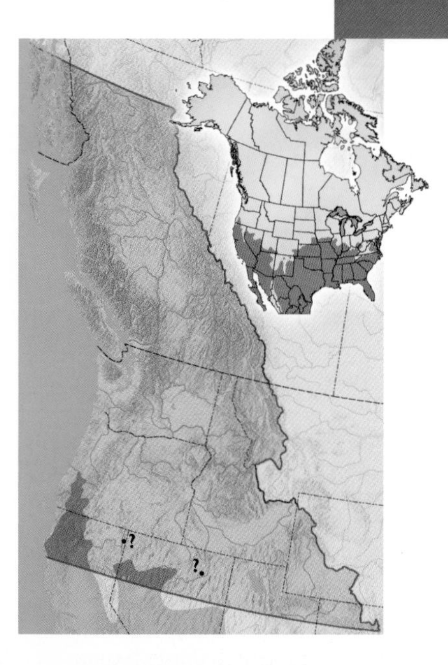

The northernmost records for Nevada are in Pershing County (at Gerlach and the Dixie Valley–Pleasant Valley area) and Lander County (in the vicinity of Battle Mountain). There are unconfirmed sightings for near Elko, Elko County, Nevada, and in extreme northeastern California in the Surprise Valley, near Eagleville, Modoc County. In the Northwest, the Common Kingsnake is found at elevations from 200 ft (60 m) along the northern California coast to 4,600 ft (1,400 m) in northern Nevada.

HABITAT AND BEHAVIOR: In northern California and the interior valleys of southwestern Oregon this reptile inhabits oak savannas, mixed pine-oak woodlands, brushy chaparral country, and the margins of farmlands. In the Great Basin, the Common Kingsnake is a desert dweller, being found in a shrub association of Big Greasewood, Shadscale, Four-winged Saltbush, and Spiny Hopsage. No

Adult; brown-banded variation that occurs in the Sacramento River drainage. Near Redding, northern California.

Common Kingsnake

Adult. Near Umpqua, Umpqua Valley, southwestern Oregon.

matter what the habitat, water is often nearby. It is largely crepuscular in its activity periods, but it may become predominantly nocturnal during hot weather. This snake can sometimes be found under rocks, logs, boards, and other surface objects. Initially, the Common Kingsnake is often defensive, coiling, striking, vibrating its tail, and attempting to bite when picked up. However, it usually becomes progressively more calm as it is handled. Another defensive tactic sometimes used is coiling into a ball with the head hidden under the body. While in this posture, the tail is elevated, with the pinkish-red lining of the vent protruded outward, and a strong-smelling musk is exuded. This reptilian predator is a constrictor that is well known for being the "enemy of the rattler." While it is true that the Common Kingsnake is immune to the venom of the rattlesnake and will occasionally eat one, it does not seek out rattlesnakes in particular. The Common Kingsnake eats several snake species, along with lizards, frogs, reptile eggs, and small mammals. This kingsnake is also a good climber and is known to sometimes ascend trees in search of the nestlings and eggs of birds. Facts about the reproduction of this species in our region are scanty. Based on records from elsewhere in its range to the south, it is probable that in

Underside of an adult. Near Hilt, Klamath River drainage, northern California.

the Northwest 2 to 12 eggs are laid during July, with hatching taking place in late August or early September.

FIELD NOTES: 24 July 1998. It was nearing 11:00 p.m. and most of my companions in the car were falling asleep. Doug Calvin, along with Tom and Casey Rodhouse and their young French friend, Francis, had met me at Pyramid Lake in northwestern Nevada to help hunt for reptiles. Although they all shared my enthusiasm for field work, their fatiguing earlier drive from Oregon was taking its toll. We had been slowly driving back and forth for over two hours on the paved highway that leads north through the parched Winnemucca Lake Basin. So far, two dead Gopher Snakes were all we had found. The average person might think that driving for hours on end at a snail's pace through the dark to look for crossing snakes would be boring. Some might even go so far as to question the sanity of anyone engaged in such a strange activity. But to the initiated who have a herpetological bent, it can be an exciting adventure on a warm desert night when conditions are ideal and snakes are on the move. Several miles may pass and anticipation builds until someone excitedly shouts, "Snake!" The brakes are applied, doors burst open, and people shoot out onto the road and sprint for the slithering animal on the pavement. Then everyone resumes the routine of peering down the yellow center line ahead, scanning for the next snake illuminated by the headlights. The radio provides background music while lively conversations and friendly debates on various subjects ensue. On this night, though, nothing was happening. But that would soon change. Just when I thought I might nod off myself, a small snake appeared in the other lane. Everyone was jarred instantly awake when they were thrown forward as my foot hit the brake pedal. We all bailed out and ran to the reptile, giving out collective "oohs" and "ahhs" when I picked up the beautiful creature. It was a Common Kingsnake with its glossy alternating bands of black and white. I knew this species inhabited the area but could not get used to the idea of it living in such an arid desert environment. I associated Common Kingsnakes with oak woodlands west of the Cascade Mountains, where the habitat is lushly vegetated and moist by comparison. But there it was in my hands; the night of road hunting was a success, and our tired spirits were renewed.

Juvenile. Near Hilt, Klamath River drainage, northern California.

Tom Rodhouse displaying living ear-jewelry—a small Common Kingsnake captured while road hunting the night before. Winnemucca Lake Basin, northwestern Nevada.

California Mountain Kingsnake

Lampropeltis zonata

Adult. Near Ashland, Rogue River Valley, southwestern Oregon.

OTHER NAMES: Coral Kingsnake.

IDENTIFICATION: Most adults are 20–30 in (51–76 cm) long, but individuals up to 40 in (1 m) are occasionally found. This beautiful reptile has a dazzling pattern of **encircling black and white crossbands with the black bands being more-or-less divided by bright red. The white bands are usually the same width on both the back and lower sides of the body.** Where the bands cross the belly, the pattern is slightly more obscure and faint. A white band extends across the top of the head, and the snout is black (occasionally with a bit of red). The dorsal scales are smooth, with a pronounced glossy quality.

VARIATION: Seven subspecies have been described, but a recent molecular genetics study indicates that some of them may be invalid. Morphologically, the races are primarily differentiated by the number of black bands that are completely split by red. Populations in the Northwest do not conveniently fit any of these subspecific categories, however, and they are currently considered to be intergrades between the SAINT HELENA MOUNTAIN KINGSNAKE (*L. z. zonata*) and the SIERRA MOUNTAIN KINGSNAKE (*L. z. multicincta*). Further taxonomic studies are needed for this species.

SIMILAR SPECIES: There are no other species of crossbanded snakes with red in their dorsal patterns that coexist with the California Mountain Kingsnake in our region.

DISTRIBUTION: In the Northwest, this snake is found at elevations from just above sea level to around 5,000 ft (1,520 m). It

occurs in most of the mountain ranges of California and through the Klamath Mountains and western slopes of the Cascade Mountains of southwestern Oregon as far north as the North Umpqua River drainage in Douglas County. The range of the California Mountain Kingsnake also extends through the Klamath River Canyon to the eastern side of the Cascade Mountains in extreme southwestern Klamath County, Oregon. It is probable that it also occurs along the eastern slopes of the Cascades in parts of northeastern California. Interestingly, there are disjunct Washington populations in the Columbia River Gorge more than 200 mi (320 km) north of the main distribution in Oregon. These are at elevations of about 100–300 ft (30–90 m) in Skamania and Klickitat counties. There are reports of this snake on the Oregon side of the Columbia River Gorge, with one questionable specimen recorded near The Dalles, Wasco County, but further confirmation is needed. Additional

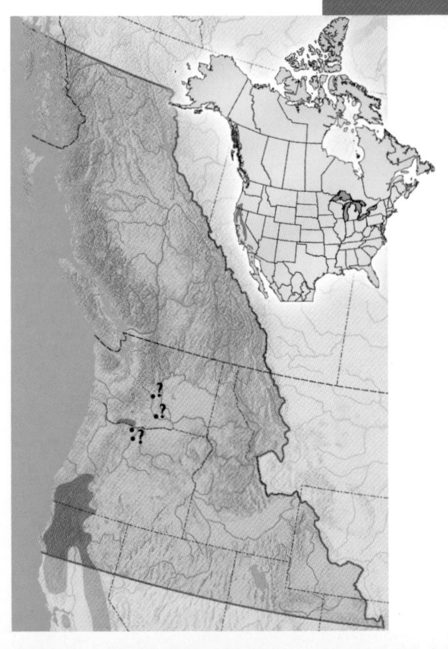

Adult. The Nature Conservancy's McCloud River Preserve, northern California.

Underside of an adult. The Nature Conservancy's McCloud River Preserve, northern California.

A young Jim Riggs and friends in 1967. Both snakes were found under the same rock, causing the gloating grin these beautiful reptiles usually elicit from naturalists. Applegate River drainage, Siskiyou Mountains, southwestern Oregon.

unconfirmed sightings come from Washington in the areas of Ellensburg, Kittitas County, and Sunnyside, Yakima County, and in Oregon at Maupin, Wasco County.

HABITAT AND BEHAVIOR: The California Mountain Kingsnake, as the name suggests, is an inhabitant of foothills and mountainous terrain. It generally avoids damp, dense coniferous forests, preferring the sunnier, more open oak and pine woodlands or brushy chaparral areas. Prime habitat for this species is a rocky canyon with a stream flowing through the bottom where there are many rock crevices and rotting logs for cover. Although diurnally active in the spring and fall, it becomes mostly nocturnal during the heat of summer. Because of its secretive nature, the California Mountain Kingsnake is not commonly seen. Anyone chancing upon this reptile is left with an indelible impression of its lustrous, colorful markings. This snake can elicit words of admiration from the most confirmed snake hater. It is usually docile when handled, only occasionally attempting to bite. Nevertheless, this lovely, harmless reptile is often killed when people mistakenly think it is a venomous coral snake. Although the patterns and coloring of both kinds of snakes are similar, coral snakes are not native to the Northwest.

Lizards, small mammals, and the eggs and nestlings of birds are food sources for this constrictor. Like other species of kingsnakes, it will occasionally eat snakes. Little is known about its reproductive habits in our region. The available data indicate that three to eight eggs are laid in June or July, with hatching taking place in August or September.

FIELD NOTES: 27 April 1993. Making a spur-of-the-moment decision that was empowered by spring fever, Gary Winter and I took a three-hour drive to the Washington side of the Columbia River Gorge. This was one of the first warm days of spring, and we speculated that reptiles should be stirring from their long winter torpor. Gary is an elementary school teacher who is extremely enthusiastic about the study of nature, and his classroom is full of terraria, aquaria, and on-going science projects. As we hiked up the canyon, I was appreciative of the help his wildlife observation skills provided. While moving through sunny openings amidst oaks and pines, looking under rocks and logs, I described how difficult it usually is to find California Mountain Kingsnakes there. Presently, we arrived at a south-sloping jumble of rocks where I knew Western Rattlesnakes, Ring-necked Snakes, and California Mountain Kingsnakes hibernated. One rattler was sunning in the opening of a crevice. I took several photographs of it, but reached the end of my film roll. While I reloaded my camera, I casually mentioned to Gary that the large flat rock he was standing near was a likely place for a Mountain King to be hiding. He immediately tilted the slab of stone upward and exclaimed, "You're correct, here's a baby one!" Picking it up, he handed me an absolutely beautiful juvenile specimen. We admired its almost gaudy markings, and I spent considerable time posing the little snake for photos before releasing it. I wish all my suggestions about where to look for rarely seen reptiles could be so successful. Gary must have profited from proverbial beginner's luck.

Gary Winter found this juvenile California Mountain Kingsnake under a sun-warmed flat rock. Columbia River Gorge National Scenic Area, Washington.

Sonoran Mountain Kingsnake

Lampropeltis pyromelana

Adult. Pine Valley Mountains, southwestern Utah.

IDENTIFICATION: Adults are usually 18–36 in (46–91 cm) long, although a few individuals slightly over 40 in (1 m) have been recorded. The Sonoran Mountain Kingsnake is often considered one of the most strikingly beautiful snakes in North America. The entire length of the body has a brilliant tricolor pattern of alternating crossbands of red, black, and white, which encircle the snake but are more pale where they cross the belly. The black bands often encroach into the red bands, splitting them dorsally at mid-back. **There is usually a distinctive narrowing of the white bands on the lower sides of the body. The neck is thin and the head prominent, elongated, and somewhat flattened. Nearly all individuals have a totally white snout, usually sharply divided from the black on top of the head by a clean, even edge.** Infrequently, there may be some black mottling within the white on the snout. The dorsal scales are smooth and glossy.

VARIATION: Several subspecies have been described as occurring in the United States and Mexico, although not all are currently recognized by some herpetologists. The Utah Mountain Kingsnake (*L. p. infralabialis*) is the race that is regarded as being found in our region.

SIMILAR SPECIES: The range of the Milk Snake (p. 202) overlaps with that of the Sonoran Mountain Kingsnake in western Utah, and the pattern and coloring of these two species are nearly identical. Milk Snakes from that area differ, however, in usually having a mostly black snout (occasional individuals have some irregular white mottling and spots on the snout), a relatively short, thick neck, and white crossbands that generally widen on the lower sides of the body. There are also differences in the number of white bands in total on the body and tail: Milk Snakes usually have fewer than 38; Sonoran

Mountain Kingsnakes generally have 43 or more. It should be noted that many Milk Snakes in Utah's Stansbury Mountains have characteristics that are similar to the Sonoran Mountain Kingsnake, such as having higher counts of white bands and extensive white markings on the snout.

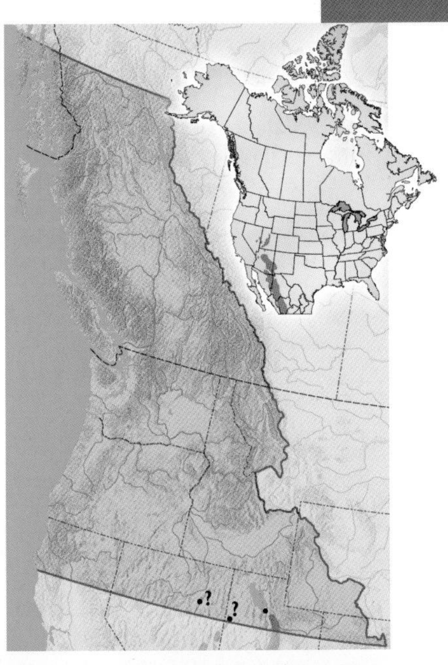

DISTRIBUTION: The Sonoran Mountain Kingsnake reaches the northernmost limits of its range within the region covered by this guide. At present, there are only four records for it in our area. Two come from the eastern slope of Utah's Wasatch Mountains, at elevations of around 6,000–7,000 ft (1,830–2,130 m), in Wasatch County. The first is a specimen listed as captured near Wallsburg in 1934 (preserved in the Brigham Young University collection); the second was found in 1998 in the southeastern foothills of Mount Timpanogos during the course of a Utah State University herpetological research project. The snake was examined closely to

Adult. Pine Valley Mountains, southwestern Utah.

make a positive identification and then released at the capture site. The other two specimens are from west of the Wasatch Mountains. One is recorded as having been collected in 1914 at Granger, Salt Lake County (preserved in the California Academy of Sciences collection). However, this community is located on the valley floor, which is unlikely habitat for this snake. The data could be erroneous and it actually may have been found in the nearby Oquirrh Mountains or some other part of the state. The fourth specimen is preserved in the Harvard Museum of Comparative Zoology and is merely listed as coming from Utah County in 1913, with no specific location given. More field surveys are needed to clarify the range of this species in northern Utah. Just to the south of our area, there are isolated populations of Sonoran Mountain Kingsnakes on some of the mountain ranges in the Great Basin of southwestern Utah and southeastern Nevada. There is a possibility that it also may be found on some of the forested mountains that rise from the arid lands of northeastern Nevada and around the perimeter of Utah's Great Salt Lake Desert.

HABITAT AND BEHAVIOR: This snake primarily inhabits mountain forests, in both the higher zones of aspen, maple, and fir, as well as the lower elevation woodlands of mixed pine and Gambel Oak. Less commonly, it has been found in the more open foothills in juniper, pinyon, and sagebrush associations. The preferred habitat seems to be rocky canyons with a streamcourse where there are ample rocks, logs, bark and other surface objects for cover. Secretive by nature, the Sonoran Mountain Kingsnake spends most of its time in hiding and is infrequently encountered crawling about. This vividly marked constrictor preys upon lizards and small rodents. Like other species of kingsnakes, it may sometimes eat snakes, but this behavior has never been documented as happening in the wild. Reportedly, the Sonoran Mountain Kingsnake is occasionally semi-arboreal when hunting for food and will eat nestling birds. Very little is known about this snake's natural history in the Northwest, especially concerning its breeding habits. Data from the southern portions of its distribution indicate that it lays three to six eggs, but there are no records revealing times of deposition and hatching at our latitude.

FIELD NOTES: 12 June 2000. More often than naturalists would prefer, excursions into the field sometimes meet with failure instead of success. I have nothing to record in my journal today regarding personal experience with the Sonoran Mountain Kingsnake in northern Utah. Although I've found it in Arizona and New Mexico, this colorful reptile has eluded me during several searches over the past four years here in the

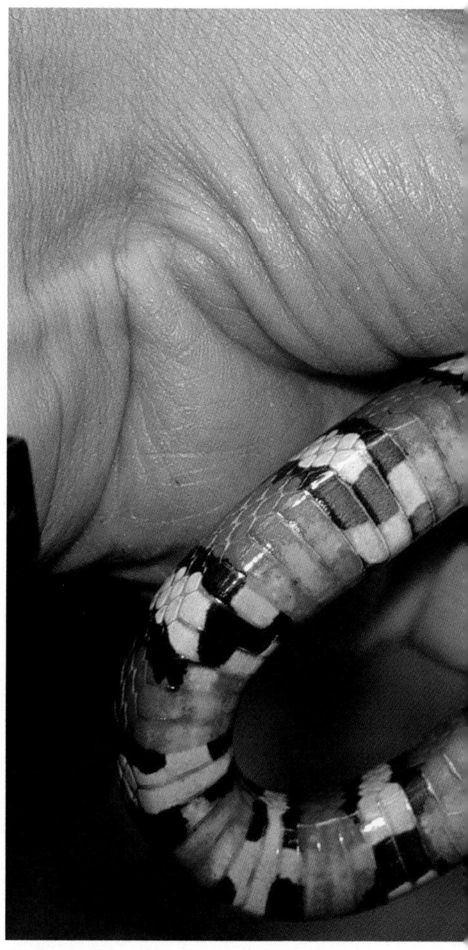

Underside of an adult. Pine Valley Mountains, southwestern Utah.

Wasatch Mountains. For the previous three days I've made a particularly intensive effort in the vicinity of where a Utah State University researcher saw one in 1998. The weather has been ideally warm, and I'd hoped to see an individual crawling across a road and photographically document it. Field studies are definitely needed to better determine the extent of this snake's distribution and habitat needs in northern Utah. After observing the general localities where the few verifiable specimens have come from, I've noticed that they are all at lower elevations in the foothills. Possibly, at the northerly extreme of its range, the Sonoran Mountain Kingsnake does not occur in high montane forests. There are also vague rumors of "coral snake" sightings in two desert-surrounded mountain ranges in the northern Great Basin that could represent this species and should be investigated. These are the Deep Creek Range on the western fringe of Utah's Great Salt Lake Desert and in the Ruby Mountains of northeastern Nevada. The Sonoran Mountain Kingsnake appears to be very rare in our region and perseverance will be required to extract its natural history facts.

Milk Snake
Lampropeltis triangulum

Adult. Foothills of Wasatch Range, near Provo, northwestern Utah.

IDENTIFICATION: Adults are 15–36 in (38–91 cm) long. This brightly marked species is extremely variable in color and pattern throughout its wide range, which extends from North America to central South America. In the western United States, Milk Snakes have a "coral snake mimic pattern" that consists of black, white, and red or orange crossbands. The black bands often encroach into the red/orange bands, splitting them dorsally at mid-back. **There is usually a distinctive widening of the white bands on the lower sides of the body.** These tricolor bands generally encircle the snake and cross the belly, but in some individuals the ventral areas are partially or completely lacking in any banding, being a uniform white. The head is black with a white band crossing at the rear. **Generally, the snout is mostly black**, but many specimens have flecks, mottlings, and spots of white or red on the snout and sometimes the top of the head. **The neck is relatively short and thick.** Dorsally, the scales are smooth and very shiny.

VARIATION: A large number of subspecies are recognized, but only one enters our area: the UTAH MILK SNAKE (*L. t. taylori*). Populations in the vicinity of two stream drainages in the Stansbury Mountains of Tooele County in northwestern Utah are very divergent, with noticeably intense markings. Individuals from that locality commonly have extensive patches of white on their snouts, and there are often areas of red on top of their heads. Many have higher counts of white bands and other characteristics that are similar to those of the Sonoran Mountain Kingsnake. Milk Snakes from east of the Wasatch Mountains in northeastern Utah and northwestern Colorado commonly have less vivid markings with pale orange crossbands.

SIMILAR SPECIES: The Sonoran Mountain Kingsnake (p. 198) coexists with the Milk Snake in western Utah, and the pattern and coloring of the two species are nearly identical. The Sonoran Mountain Kingsnake differs, however, in nearly always having a completely white snout, usually sharply divided from the black on the top of the head by a clean, even edge (occasional individuals have black mottling within the white). The neck is thinner, the head is longer and slightly flattened, and the white crossbands become narrower on the lower sides of the body. Additionally, the Sonoran Mountain Kingsnake has 43 or more white crossbands in total on the body and tail, whereas the Milk Snake usually has fewer than 38. The Long-nosed Snake (p. 206) also resembles the Milk Snake, and both are found in proximity in western Utah. They can be easily differentiated by the caudal scales on the underside of the tail. Most of the Long-nosed Snake's caudals are in one undivided row, whereas

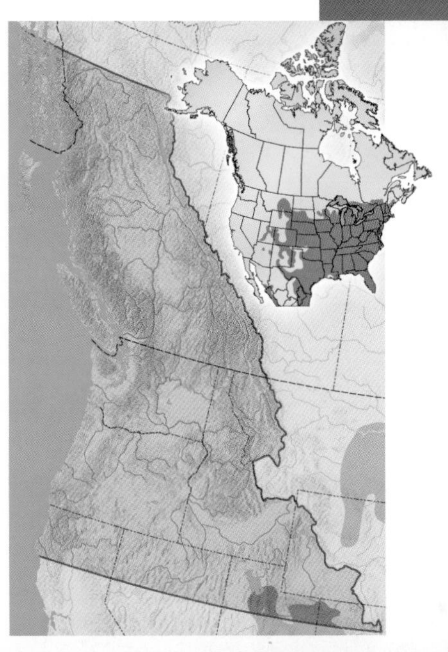

Adult. Stansbury Mountains, northwestern Utah.

Underside of an adult with a banded belly. Stansbury Mountains, northwestern Utah.

Underside of an adult with a uniformly white belly. Stansbury Mountains, northwestern Utah.

Adult with white mottling on the snout and patches of red on the top of the head, typical of many individuals from the Stansbury Mountains of northwestern Utah.

the Milk Snake has caudals that are divided in the middle and arranged in two parallel rows.

DISTRIBUTION: This reptile enters our region only in northern Utah and northwestern Colorado, at elevations of 4,000 ft (1,220 m) to just over 6,000 ft (1,830 m). In northwestern Utah, the Milk Snake has been recorded from the area between the Great Salt Lake Desert and the western face of the Wasatch Mountains, as far north as the Salt Lake City area. East of the Wasatch Mountains the Milk Snake is found in Utah's Uinta Basin and in the Yampa River drainage of adjacent Colorado.

HABITAT AND BEHAVIOR: The Milk Snake occurs from grassy savannas and sagebrush-covered open valleys and lower foothills upward through mixed juniper and pinyon woodlands into mountain canyons that have a growth of Gambel Oak, Canyon Maple, and pine forests. It avoids deserts. Although this snake may be encountered at any time of the day, during hot weather it becomes largely nocturnal. It is secretive and spends most of its time hiding under rocks, logs, bark, and other surface litter, or within subterranean rodent burrows. In eastern North America, it is common in farmland areas and is often seen around barns. Apparently, this behavior has somehow led to the bizarre belief that this snake sucks milk from cows, resulting in the common name. Instead of stealing milk, these snakes frequent barns to

eat mice, and they actually aid farmers in rodent control. Additional components of the diet include other small mammals, lizards, snakes, and birds. The eggs of reptiles and birds are also occasionally eaten. Constriction is used to subdue prey. Although normally gentle when handled, the Milk Snake will sometimes suddenly begin chewing on its captor's hand in an industriously prolonged manner. Our knowledge of this species' breeding habits in the Northwest is meager. Records from elsewhere in the American West indicate that 2 to 17 eggs are deposited during July and hatch in late August or early September.

FIELD NOTES: 25 August 1998. This was my fourth night of roadhunting for Milk Snakes in a wide canyon that opens out of the Wasatch Mountains near Provo, Utah. Conditions were very dark and windy, with the air temperature rapidly cooling. Local herpetologists had urged me to try this particular road, painting tantalizing word pictures of their past successes in finding this rare reptile there. After six passes up and down the canyon, it seemed as though my weary eyes had memorized every detail on the paved surface. I had already taken photos of two specimens of this species in the Stansbury Mountains, but Milk Snakes from that particular area are quite unusual in appearance, with beautifully bright markings and white snouts. Most Utah Milk Snakes have snouts that are primarily black. Although it is always tempting to photograph exceptional individuals, it is better to illustrate an identification-oriented field guide with the more typical-looking examples of each kind of reptile. After two weeks of searching, "ordinary" Milk Snakes were still eluding me. I knew this was my last chance, as I'd be leaving the area the next day. At about 9:15 p.m., the headlights lit up a snake on the road. It appeared to be about 20 in (50 cm) long and was frantically attempting to wriggle off the road, almost being aided in the effort by gusts of wind. It was a Milk Snake with a more typical pattern and coloration on the body. Nevertheless, my elated, self-congratulatory mood quickly ended when I shined my flashlight on the snake's head. It had a reddish-brown snout! I still didn't have a black-snouted Utah Milk Snake.

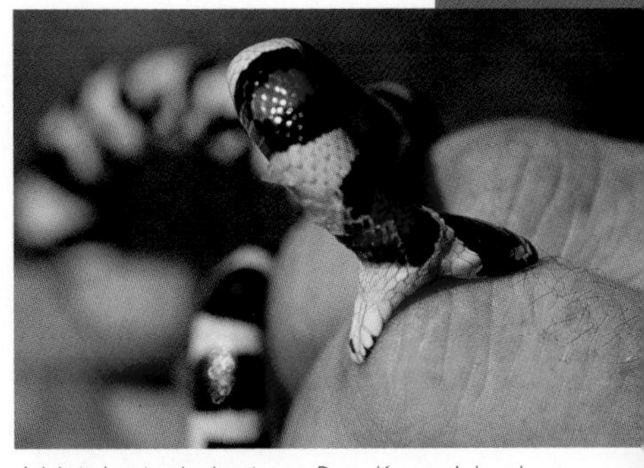

Adult, industriously chewing on Doug Knutsen's hand. Stansbury Mountains, northwestern Utah.

Habitat of the Milk Snake in the western foothills of the Wasatch Range, northwestern Utah.

Long-nosed Snake
Rhinocheilus lecontei

Adult. Bruneau River Canyon, southwestern Idaho.

IDENTIFICATION: Adults are 20–40 in (51–102 cm) long. Like its close relatives, the kingsnakes, this rather slim-bodied snake is brightly crossbanded. It differs in that **the overall dorsal pattern has a distinctive speckled appearance.** There is a series of broad, black bands on the back that tend to fade into white or cream speckles on the lower sides of the body. The interspaces between the black bands are white or cream colored, usually with orangish-red speckles at mid-back, grading to black speckles along the sides of the body (exclusive of a narrow white or cream gap that borders each black band). In some individuals the red areas may be more solid and intense; in others the red may be reduced to only a faint pinkish tinge. Occasionally, the red coloring is entirely absent, and the snake has a totally black-and-white pattern. The top of the head is solid black at the rear but is marked with white or cream toward the snout and on the lip scales. As the common name suggests, **the snout is elongated and pointed, with a shorter, countersunk lower jaw. An enlarged rostral scale curves upward over the tip of the snout.** The dorsal scales on the body are smooth. This snake is the only colubrid in the American West that has **most of the caudal scales on the underside of the tail arranged in a single, undivided row.** The dorsal crossbands do not continue onto the ventral surfaces, the belly being a uniform white or cream, sometimes with a few dark mottlings. Juveniles often have only faint speckles along the sides of the body or lack these markings entirely.

VARIATION: Two subspecies are found in the United States, with others in Mexico. One occurs in our region: the WESTERN LONG-NOSED SNAKE (*R. l. lecontei*).

SIMILAR SPECIES: The Common Kingsnake (p. 190), the Milk Snake (p. 202), and

two variations of the Ground Snake (p. 236) have similarly vivid dorsal crossbanding, and their ranges overlap with the Long-nosed Snake in parts of the Northwest. However, the Long-nosed Snake is readily differentiated from them by its mostly single, undivided row of caudal scales on the underside of the tail.

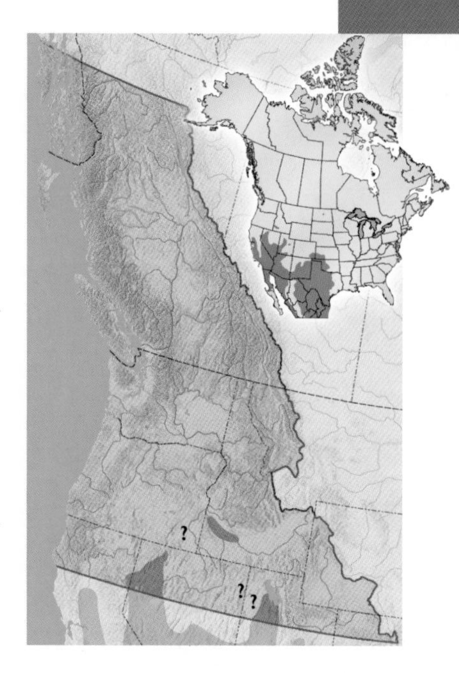

DISTRIBUTION: The Long-nosed Snake ranges northward through the Great Basin of Nevada, western Utah, and northeastern California (Honey Lake Basin). There are also disjunct populations in the Snake River Valley of southwestern Idaho. The known northern limits of this reptile's distribution in North America extend very close to the borders of southeastern Oregon. It has been recorded in northwestern Nevada south of Denio, Humboldt County (eastern foothills of the Pine Forest Range), and in southwestern Idaho's Canyon County (a few miles northwest of Murphy.) It is probable that future field surveys will find this species in Harney or Malheur Counties, Oregon. There is suitable habitat for the Long-nosed Snake in parts of northeastern Nevada and it may someday be found there, along with the western portions of Utah's Great Salt Lake Desert. Elevations for this species in the Northwest vary from around 2,300 ft (700 m) in the Snake River drainage of Idaho to nearly 4,500 ft (1,370 m) in the Black Rock Desert of Nevada.

HABITAT AND BEHAVIOR: In our region, this snake is an inhabitant of deserts in the Great Basin shrub communities of

Adult with reduced red in pattern. Winnemucca Lake Basin, northwestern Nevada.

Adult. Smoke Creek Desert, northwestern Nevada.

open flats and rocky canyons. Typical plants in its habitat are Big Greasewood, Shadscale, Four-winged Saltbush, and Spiny Hopsage. Equipped with a pointed snout and countersunk lower jaw, the Long-nosed Snake is well adapted for burrowing, and it usually occurs in areas where the soil is loose and sandy. It hides under rocks or in subterranean burrows during the day and is seldom seen. Because the Long-nosed Snake has crepuscular and nocturnal activity periods, the only time it is usually encountered is when one is observed crossing a road on a warm night. When handled, this reptile will usually nervously vibrate the tail, but it rarely bites. Occasionally, it exhibits a defensive display of

Adult, showing the elongated snout and countersunk lower jaw. Bruneau River Canyon, southwestern Idaho.

repeatedly coiling and uncoiling in a writhing manner. The head is hidden beneath the body, along with elevating the tail and discharging bloody feces and foul-smelling musk from the vent. Important food sources for the Long-nosed Snake are other reptile's eggs, which it finds through its burrowing activities, and lizards that sleep in the sand at night. Small mammals, insects, and other snakes are included in the diet, as well. Constriction is used to kill larger prey. In the American Southwest, 4 to 11 eggs are deposited during June or July and hatch in August. Although there are no data about the reproductive habits for this species in the Northwest, it is probable that at our more northerly latitude the eggs are laid in mid- to late July, with hatching taking place at the end of August or in early September.

FIELD NOTES: 8 July 1998. Although the night had been warm, no snakes had turned up on the roads I had cruised around Bruneau Dunes State Park in southwestern Idaho. The sandy habitat there is perfect for Long-nosed Snakes, but at 1:00 a.m. I gave up, drove back to camp, and crawled into my sleeping bag. While gazing at the starry desert sky, I enjoyed the memories of a more successful trip here with Jim Riggs in August of 1973. At that time, I needed photographs of a Long-nosed Snake and had been periodically searching for one in the Northwest for nearly six years. We felt lucky on the warm evening we arrived and immediately began hunting on some ideal-looking blacktopped roads near the dunes. However, after patrolling the roads until about midnight, all we had seen were two Gopher Snakes. Because it now seemed too cool for reptile activity, we drove back to our camp, situated several miles up the Bruneau River Canyon. We were only a few blinks and nods away from sleep, and just a short distance from camp, when something elongated appeared in the rutted road. Both of us saw it, but was it a snake or a stick? We hurried out of the car to examine it and in unison exclaimed, "Long-nosed Snake!" In our haste to capture the creature, we quickly bent over it and bumped our heads together with a painful whack. Despite hours of earlier road hunting on the warm, "ideal" blacktop, we had at last found our rare quarry on a narrow dirt track only a few paces from our camp, at 12:45 a.m. in the cold. I later learned that Long-nosed Snakes can tolerate fairly low temperatures. After years of searching, I had finally met with success in a place and time of night I had least expected. As one herpetological sage put it, "Snakes are where you find them."

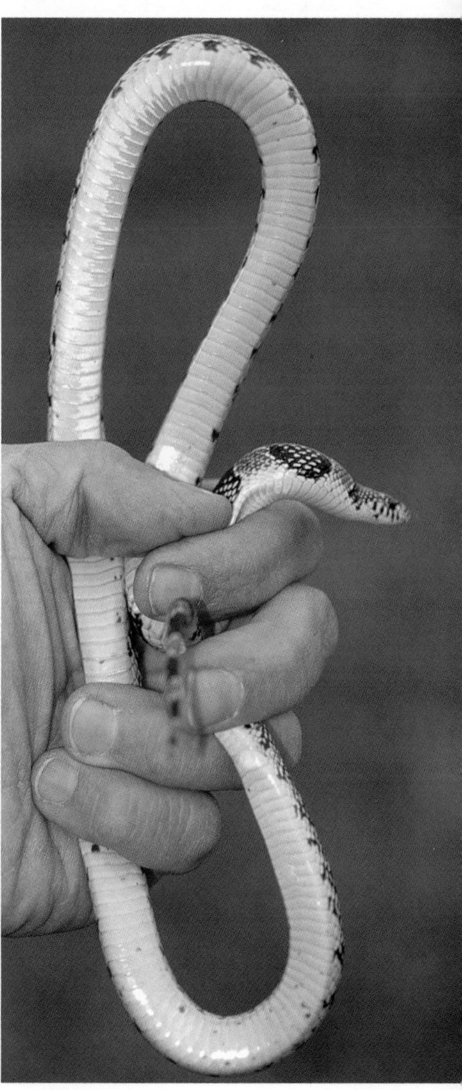

Underside of adult. Smoke Creek Desert, northwestern Nevada.

Common Garter Snake

Thamnophis sirtalis

Adult Valley Garter Snake. Deschutes River Canyon, near Madras, central Oregon.

IDENTIFICATION: Adults average 18–36 in (46–91 cm) long, but specimens of up to 52 in (1.3 m) have been recorded. **There are usually seven upper labials** (lip scales) on each side of the head, and, as in all other garter snakes, the dorsal scales on the body are strongly keeled. The coloration and pattern of the Common Garter Snake is variable, both geographically and sometimes within a single population. Except for one subspecies, **most individuals have an extremely bright, wide, light dorsal stripe running the entire length of the mid-back.** There is often a pale to fairly well-defined, lengthwise stripe along each side of the body, as well. The stripes may be yellow, greenish yellow, turquoise, or white. The ground color between the stripes is usually black or dark gray, **often with small to large blotches of red.** Throughout most of the Northwest, the top of the head is black, but in some variations the head may be partially or entirely red. Ventrally, the coloring on the throat is shades of yellow, blue, or green, grading to black on the belly and tail.

VARIATION: This snake is the most widespread reptile in North America, and there is a high degree of diversity in its physical appearance from coast to coast. Depending upon the prevailing taxonomic opinion, 11 or 12 total subspecies are recognized, with 5 occurring in our region. The RED-SPOTTED GARTER SNAKE (*T. s. concinnus*) has a wide, yellow or greenish-yellow dorsal stripe, but it usually lacks side stripes. The sides of the body are black, with large red or orangish-red blotches. The entire head is bright red or orangish red. One uncommon color morph lacks the red markings, instead having a white or bluish-white dorsal stripe, side blotches, and head. This subspecies occurs in northwestern Oregon, southwestern Washington, and the coastal portions of the Olympic Peninsula. The most typical examples are

found in Oregon's Willamette Valley, whereas those from the surrounding areas often have some characteristics of the *fitchi* subspecies. The PUGET SOUND GARTER SNAKE (*T. s. pickeringii*) has a very narrow, turquoise or greenish-yellow dorsal stripe and side stripes of the same color. The area between the stripes is black, usually with light, narrow bars and flecks that match the color of the stripes or are occasionally pale orangish red. The top of the head is black, and the sides of the head and the lip scales are pale turquoise or greenish yellow. It occurs in the Puget Lowlands of Washington and British Columbia, including Vancouver Island. The CALIFORNIA RED-SIDED GARTER SNAKE (*T. s. infernalis*) has a wide, yellow, greenish-yellow, or bluish-white dorsal stripe with equally or wider side stripes (sometimes extending down to and merging with the light belly). The area between the stripes is black, with large red blotches. The top and sides of the head are red, and the lip scales are white. It occurs on the northern California coast south of the Klamath River. The VALLEY GARTER SNAKE (*T. s. fitchi*) has a wide, yellow or greenish-yellow dorsal stripe, along with a similarly colored, paler stripe along each side of body. The area between the stripes is black, with small red blotches. The top of the head is black (sometimes with red on the sides). There are no black spots on the ends of the ventral scales along both edges of the belly. Individuals from the shores of Crater Lake, Oregon, are uniformly dark gray and lack the typical markings. This subspecies is the most wide-ranging one in the Northwest, occurring throughout much of the intermountain region east of the Cascades (except for the Great Basin Desert). It is also found in southwestern Oregon, northern California, mainland southern British Columbia (exclusive of the southwestern coast and far eastern portions), and the extreme southeastern tip of Alaska. The RED-SIDED GARTER SNAKE (*T. s. parietalis*), which is very similar in appearance to the Valley Garter Snake, has a typical, wide, yellow or greenish-yellow dorsal stripe, but the stripes on the sides are frequently orangish yellow, sometimes with a reddish tinge. The area between the stripes has a pattern of large

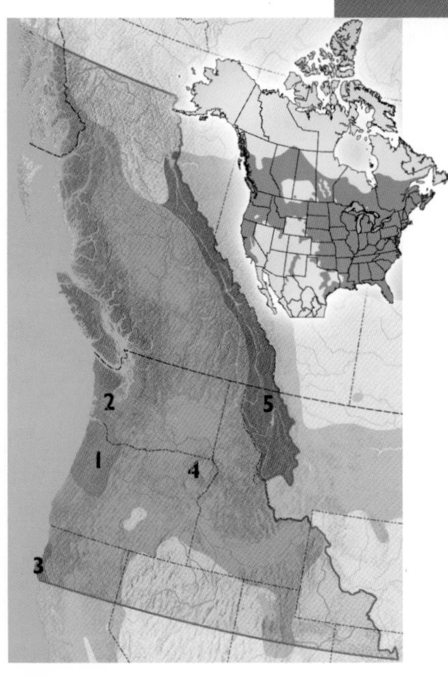

SUBSPECIES

- **1** Red-spotted Garter Snake *T. s. concinnus*
- **2** Puget Sound Garter Snake *T. s. pickeringii*
- **3** California Red-sided Garter Snake *T. s. infernalis*
- **4** Valley Garter Snake *T. s. fitchi*
- **5** Red-sided Garter Snake *T. s. parietalis*
- Species range outside the Pacific Northwest

Adult Red-spotted Garter Snake. Wilson Wildlife Area, near Corvallis, Willamette Valley, northwestern Oregon.

Common Garter Snake

Adult California Red-sided Garter Snake. Near Eureka, northern California coast.

black spots interspersed with red or orangish-red areas that may form bars or be reduced to a red suffusion between the scales. The top of the head is often olive colored. There is a black spot on either end of each ventral scale along both edges of the belly. The Red-sided Garter Snake occurs in our region only in the Rocky Mountains area of extreme eastern British Columbia and western Montana. It intergrades with the Valley Garter Snake in some locations, where snakes with characteristics of both subspecies may be encountered. The Common Garter Snake is especially variable in western Montana: many individuals appear to be typical of the *fitchi* subspecies; others have a solid black ground color between the stripes, with no red in the pattern, and they resemble neither the *parietalis* nor *fitchi* subspecies.

Adult Puget Sound Garter Snake. Near Olympia, western Washington.

SIMILAR SPECIES: All five species of garter snakes native to the Northwest have similar patterns of lengthwise stripes and can be difficult to differentiate. If in doubt, use the Quick Key to Garter Snakes (pp. 68–69). The Striped Whipsnake (p. 166), California Whipsnake (p. 170), and Western Patch-nosed Snake (p. 178) also have striped patterns, but they are easily identified by their large, smooth dorsal scales. All garter snakes have strongly keeled scales.

DISTRIBUTION: Within the region covered by this guide, the Common Garter Snake is found nearly everywhere, from sea level to just over 7,000 ft (2,130 m). It is the only reptile known to occur in Alaska (on Kupreanof Island and in the Stikine and Taku river drainages). It is absent from the cold northern

Adult Red-spotted Garter Snake. Wilson Wildlife Area, near Corvallis, Willamette Valley, northwestern Oregon.

sections of the Northwest, high alpine zones on mountain ranges, and the deserts of the Great Basin. It also appears to be absent from parts of western Wyoming, northeastern Utah, and northwestern Colorado.

HABITAT AND BEHAVIOR: This snake is usually encountered around aquatic habitats. Occasionally, it will stray fairly far from water in the moist forests and meadows west of the Cascade Mountains. In the dry interior plateau, it is associated with the vegetated riparian zones of creeks, rivers, lakes, and marshes. Primarily diurnal during warm weather, Common Garter Snakes are often seen along the edges of ponds and streams hunting for food or basking on floating logs and cattail mats. When alarmed, they will

Underside of a Red-spotted Garter Snake. Near McMinnville, Willamette Valley, northwestern Oregon.

Juvenile Red-spotted Garter Snakes. Wilson Wildlife Area, near Corvallis, Willamette Valley, northwestern Oregon.

invariably escape into the water, often diving to the bottom and remaining submerged until danger has passed. They can also frequently be found hiding under surface objects, such as logs, bark, rocks, and boards. If threatened, they might vigorously strike and bite in defense. In typical garter snake fashion, they will emit a foul-smelling musk from glands by the vent and release feces and excrement on the hands of their captor. They sometimes travel considerable distances to reach rocky hibernation sites on south-facing slopes, where large numbers may gather. Hibernacula are often shared with other species of garter snakes, Racers, Rubber Boas, and Western Rattlesnakes. Studies have shown that juveniles eat mainly earthworms, whereas adults largely feed upon such aquatic prey as small fish and amphibians and their larvae. Small mammals and birds are also sometimes eaten. Some populations of the Common Garter Snake are immune to the toxic skin secretions of the Western Toad (*Bufo boreas*) and newts of the *Taricha* genus and include those amphibians in their diet. Although it is more common for 3 to 18 young to be live-born in late July or during August, the Common Garter Snake is known to occasionally produce as many as 85 young in one birthing.

FIELD NOTES: 28 May 1984. While investigating the edges of an old mill pond near Fairdale in the Coast Range foothills of Yamhill County, Oregon, I chanced upon a Common Garter Snake. Carefully parting some cattails to obtain an unobstructed view, I saw that it was in the process of swallowing a Rough-skinned Newt (*Taricha granulosa*). This scene grabbed my total attention, because the Common Garter Snake is one of the few predators that can eat this extremely toxic amphibian and survive. Research has shown that this snake is 2,000 times less susceptible to poisoning from the skin secretions of the newt than is a white lab mouse. This particular individual, however, did not seem totally immune to all effects of the toxin. As I continued to observe this fascinating drama in the pond's shallows, I noticed that the snake was behaving in a rather odd manner. Its movements were becoming increasingly sluggish and less effective as the salamander was gradually engulfed. After the tip of the tail finally disappeared down the snake's throat, the snake opened and closed its jaws in a rather dazed manner and then just floated motionless in the water. After a few minutes, I reached out and easily captured the snake. Hanging limply from my grasp, the reptile made only feeble attempts to wriggle out of my grip. It appeared to be inebriated. I released it on the grassy bank of the pond and it slowly disappeared down a hole. I couldn't help but muse about whether or not the snake would have a horrendous hangover after sleeping off its bout of "drunkenness."

Red-spotted Garter Snake eating a Rough-skinned Newt (Taricha granulosa). *Near McMinnville, Willamette Valley, northwestern Oregon.*

Northwestern Garter Snake

Thamnophis ordinoides

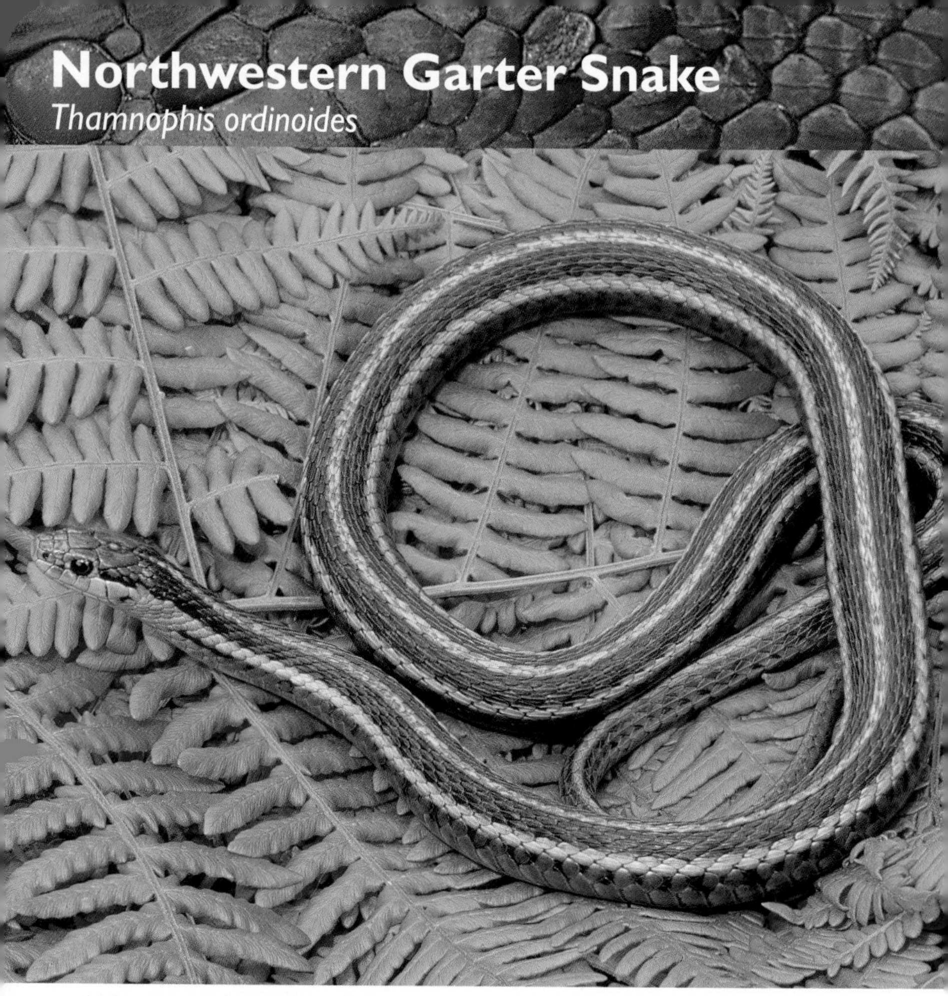

Adult; orange-striped morph. Eugene, Willamette Valley, northwestern Oregon.

IDENTIFICATION: This garter snake is small, with most adults averaging 12–24 in (30–61 cm) long, although occasional individuals slightly exceed 36 in (91 cm). **There are usually seven upper labials (lip scales) on each side of the head.** Typical of a garter snake, the dorsal scales on the body are strongly keeled. **The head is relatively small in size when compared to the other garter snakes found in our area.** This garter snake also bears the distinction of being the most variable in color and pattern of any snake in the region. Although the dorsal stripe down the middle of the back is sometimes very narrow and dull or entirely absent, on most individuals it is wide and well defined. There may or may not be a stripe along each side of the body. The color of the dorsal stripe can range from bright yellow or red, through orange, greenish yellow, turquoise, blue, white, or dull yellowish tan. When present, the side stripes are usually yellow, tan, cream, or white. The ground color between the stripes may be black, brown, olive, or gray, often with two rows of alternating dark spots, and sometimes with white flecking. The belly is generally brown, gray, black, or yellowish, often with black markings. Some red-striped morphs will have bright red flecks scattered throughout the dorsal ground color, side stripes, and on the belly. In addition, there may be a reddish cast to the snake's entire ground color. Similarly, Northwestern Garter Snakes with bluish or greenish dorsal stripes will often have an overall suffusion of these colors. Entirely red

specimens without stripes have been found in the Siskiyou Mountains of extreme northwestern California. Melanisitic individuals that are completely black, with little or no apparent markings, occasionally occur throughout this species' range. All or several of the above-described morphs can be found in a single population, and two or three of these variations may be produced in a single litter of young.

VARIATION: Although some geographic populations may have one morph that tends to predominate (e.g., more of the red-striped variations seem to occur in the southern portions of this species' range), no subspecies are recognized.

SIMILAR SPECIES: The five species of garter snakes native to the Northwest have similar patterns of lengthwise stripes and can be difficult to differentiate. If in doubt, use the Quick Key to Garter Snakes (pp. 68–69). The Striped Whipsnake (p. 166), California Whipsnake (p. 170), and Western Patch-nosed Snake (p. 178) also have striped patterns, but they are easily identified by their large, smooth dorsal scales. All garter snakes have strongly keeled scales.

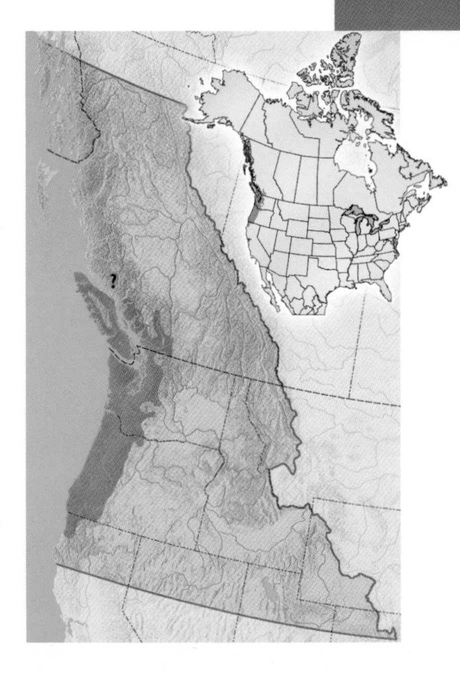

DISTRIBUTION: This snake is well named, because the entirety of its range lies within the Northwest region, at elevations from sea level to around 5,500 ft (1,680 m). It occurs west of the Cascade Mountains from

Adult; red-striped morph. Eola Hills, Willamette Valley, northwestern Oregon.

Adult; reddish orange–striped morph. Near Tumwater, western Washington.

Adult; yellow-striped morph. McMinnville, Willamette Valley, northwestern Oregon.

Adult; yellowish-tan, single-striped morph. Near Corvallis, Willamette Valley, northwestern Oregon.

the extreme northwestern coastal corner of California (Del Norte County), north through western Oregon and Washington, to southwestern British Columbia (including Vancouver Island and many of the smaller adjacent islands). The range of the Northwestern Garter Snake extends barely east of the Cascades through the Columbia River Gorge and over the Snoqualmie and White passes in Washington.

HABITAT AND BEHAVIOR: This snake lives in damp, heavily vegetated habitats of the Northwest and is one of the few reptiles able to thrive in the foggy coastal sections of the region. Some sunshine is needed, however, and it is rarely found in dense, shady forests. Most commonly, it is seen in grassy-brushy places at the edges of forests, in meadows, along roadside ditches and embankments, and in clearings where the forest canopy has been opened by natural disturbance or logging. Northwestern Garter Snakes are often encountered around old homesteads that have scattered boards and other litter for cover, and in urban settings where there are vacant lots and weedy sections of backyards. Because it is often seen around gardens, it is sometimes called a "gardener snake," whereas the red-striped morph is incorrectly referred to as a "red racer." Although occasionally found near water, this species is chiefly terrestrial. Northwestern Garter Snakes have been observed hunting for prey in dripping vegetation during summer rainstorms. Befitting a snake of such moist environments, slugs and earthworms comprise most of the diet. Snails and small amphibians have also been reported as being eaten at times. When handled, it rarely resorts to biting, but like other garter snakes, feces and a foul-smelling musk are emitted from the vent to repel a would-be captor. Interestingly, studies have shown that striped morphs of this species flee when threatened—a striped pattern visually confuses a predator about the speed and direction of the moving snake—whereas spotted morphs with faint stripes remain motionless and try to avoid detection by depending upon their patterns to camouflage them in the habitat. Three to 20 young are live-born during July, August, or September.

Northwestern Garter Snake

Adult. Near Tumwater, western Washington.

FIELD NOTES: 25 December 1977, 2:30 p.m. I had never made field observations of native Northwestern reptiles on Christmas Day before, but there's a first time for everything. While celebrating with the family at my parents' home in Brookings on the southern Oregon coast, I went outside with my two-year-old son, Shawn. We wanted to enjoy the surprisingly warm sunshine and the sweeping view of the Pacific Ocean. The climate in this part of the state is noted for its mild winters and is sometimes humorously called "Oregon's banana belt." This particular day was living up to that reputation. While soaking up the sunny warmth, we soon discovered that we weren't the only ones basking in the solar rays. Shawn, who was prowling around at ground level on an embankment, suddenly exclaimed, "Nake, Daddy! Nake!" He was pointing at a Northwestern Garter Snake that was coiled on a clump of grass. Because the reptile was barely warm enough to be active, I easily captured it and gave my youngster the opportunity to touch the scales and examine its beautiful red dorsal stripe. He was delighted with the slithering creature, but he wrinkled up his nose when he got a whiff of the frightened snake's repulsive-smelling musk. We found two more Northwestern Garter Snakes in the immediate vicinity. It is actually fairly common for this species to briefly emerge from hibernation on warmer winter days in the low-elevation areas of the Northwest. Since that time, I have seen these snakes out sunning on several occasions during balmy winter periods in Oregon's Willamette Valley, but that unique Christmas surprise in 1977 still remains special in my memory.

Underside of adult. Eugene, Willamette Valley, northwestern Oregon.

Juvenile. Near Victoria, Vancouver Island, British Columbia.

Western Terrestrial Garter Snake

Thamnophis elegans

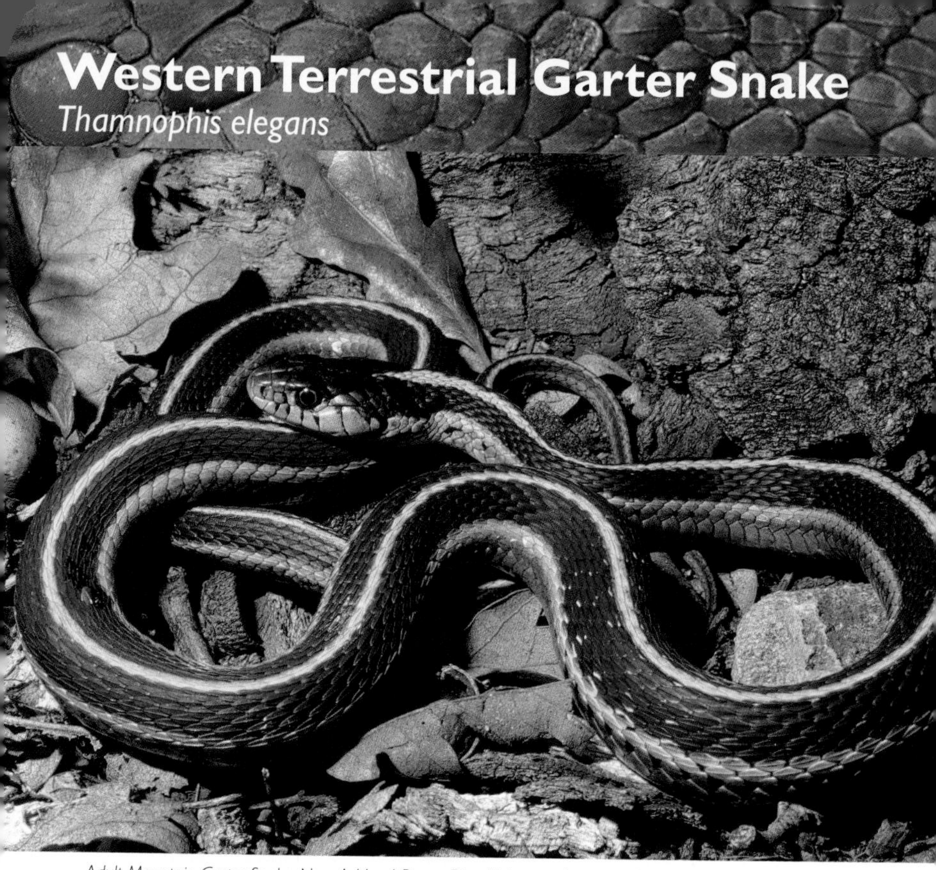

Adult Mountain Garter Snake. Near Ashland, Rogue River Valley, southwestern Oregon.

IDENTIFICATION: Adults are 18–43 in (46–109 cm) long. This species **usually has eight upper labials (lip scales) on each side of the head. The internasal scales on top of the snout are wider than long and are not pointed at the front end.** As in all garter snakes, the dorsal scales on the body are strongly keeled, and a light dorsal stripe usually runs down the middle of the back, with a matching stripe along each side of the body. Depending upon the geographic population, the color, intensity, and definition of these stripes is very diverse. The ground color between the stripes is also variable, ranging from solid black through light or dark shades of gray, brown, or reddish brown. There is usually a pattern of small or large black spots within the ground color. The belly is pale gray, dark gray, bluish, greenish, or brown, often with black mottlings and flecks (red flecks in the Coast Garter Snake subspecies). Because of this confusing array of coloring and pattern, it is important to refer to the descriptions of the subspecies when making an identification.

VARIATION: Of the six subspecies currently recognized, three occur in the Northwest. The MOUNTAIN GARTER SNAKE (*T. e. elegans*) has a broad, bright yellow or orangish-yellow dorsal stripe with straight, even edges and a well-defined, pale yellow stripe along each side of the body. The color between the stripes is usually solid black, sometimes with small light flecks. It occurs throughout the interior of northern California and western Oregon. Willamette Valley populations differ in having an irregular-edged, bright white (occasionally pale yellow) dorsal stripe that has lengthy sections broken into discontinuous spots or diamond shapes. The side stripes are replaced by a bluish-gray coloration that extends upward from the belly onto the sides, where it merges between a pattern of large black spots. The COAST GARTER SNAKE (*T. e. terrestris*) is similar to the Mountain

Garter Snake subspecies, but the color between the stripes is brown or olive, with a pattern of dark spots, intermixed with a suffusion of rusty-red coloring. There are usually red flecks on the side stripes and belly, as well. It occurs along the coasts of northern California and extreme southwestern Oregon. The WANDERING GARTER SNAKE (*T. e. vagrans*) has a dull yellow dorsal stripe that is narrow (sometimes evident only on the neck) and usually being encroached upon by dark spots, creating an uneven, "zig-zag" appearance. The side stripes are a more dull shade of yellow, and the ground color of the body is light gray or brown, with a pattern of well-separated, small, dark spots. It occurs throughout most of the Northwest east of the Cascade Mountains, as well as in western Washington and southwestern British Columbia. Many individuals from the Puget area have a darker ground color, and those populations were once considered a separate subspecies (*T. e. nigrescens*). Melanistic (all black) individuals with no discernible stripes occur within some normally patterned populations throughout the range of this subspecies. Western Terrestrial Garter Snakes in Klamath and Lake counties of south-central Oregon and Siskiyou and Modoc counties of northeastern California are larger and more heavy-bodied. Their striped patterns and coloring have characteristics of both the *vagrans* and *elegans* subspecies. Although some herpetologists consider them to be intergrades between these two forms, others favor classifying them

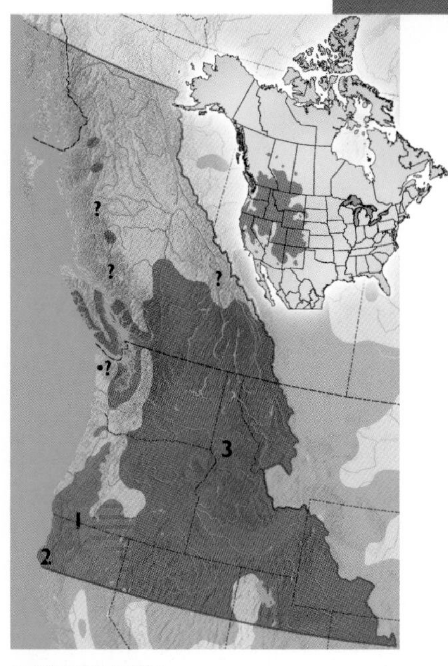

SUBSPECIES

1	Mountain Garter Snake *T. e. elegans*
2	Coast Garter Snake *T. e. terrestris*
3	Wandering Garter Snake *T. e. vagrans*
	Area of intergradation (*biscutatus* form)
	Species range outside the Pacific Northwest

Adult Mountain Garter Snake; Willamette Valley variation. Flying M Ranch, North Yamhill River, northwestern Oregon.

Adult Wandering Garter Snake; light variation. Wasatch Range, southeastern Idaho.

Adult Coast Garter Snake. Near Brookings, southern Oregon coast.

as a distinct subspecies, the KLAMATH GARTER SNAKE (*T. e. biscutatus*). Additionally, some authorities recognize another subspecies as occurring in the Colorado River drainage of eastern Utah: the UPPER BASIN GARTER SNAKE (*T. e. vascotanneri*), which has only a faint dorsal stripe that is broken into dull yellow spots or may be entirely absent. The side stripes are pale yellow and poorly defined, and the ground color of the body between the stripes is olive-brown or grayish tan, with dark bars and spots.

SIMILAR SPECIES: The five species of garter snakes native to the Northwest have similar patterns of lengthwise stripes and can be difficult to differentiate. If in doubt, use the Quick Key to Garter Snakes (pp. 68–69). The Striped Whipsnake (p. 166), California Whipsnake (p. 170), and Western Patch-nosed Snake (p. 178) also have striped patterns, but they are easily identified by their large, smooth dorsal scales. All garter snakes have strongly keeled scales.

DISTRIBUTION: The Western Terrestrial Garter Snake ranges throughout the greater portion of the Northwest, from sea level to nearly 9,000 ft (2,740 m). It is often the most commonly encountered garter snake in the interior plateau country, and it is found eastward from the Cascade Mountains to the Rocky Mountains, and north from Nevada,

Utah, and Colorado to southern British Columbia. West of the Cascades, this snake occurs from California northward through southwestern Oregon and into the Willamette Valley along the western foothills of the Cascade Mountains, where it has been recorded as far north as Molalla, Clackamas County. There are also isolated populations on the northwestern side of the Willamette Valley in the foothills of the Coast Range in Yamhill, Washington, and Columbia counties. In western Washington and southwestern British Columbia, it has a spotty distribution throughout the Puget Sound area, including Vancouver Island and many of the smaller islands. There are apparently disjunct coastal populations of this species in British Columbia's Bella Coola, Skeena, and Nass river drainages. It is possible that future field surveys will find that this reptile has a continuous distribution along the British Columbia coast, possibly as far north as the extreme southeastern tip of Alaska. There is a single 1954 specimen from the northwestern coast of Washington's Olympic Peninsula in Clallum County (preserved in the University of Michigan collection). This record has never been duplicated and the locality data may be erroneous.

HABITAT AND BEHAVIOR: The Western Terrestrial Garter Snake inhabits a variety of ecosystems and elevations. It is

Adult Wandering Garter Snake; dark nigrescens *variation. Near Olympia, western Washington.*

Adult Wandering Garter Snake; black melanistic individual. Near Rome, Owyhee River Canyon, southeastern Oregon.

found in the more open sections of coniferous forests along parts of the Northwest's coasts and on islands in the Puget area. Inland, this snake occurs in the oak savannas and brushy chaparral of valleys, foothills, and mountains, and ranges throughout much of the Intermountain West in dry pine woodlands and semi-arid juniper-sagebrush associations. On some mountain ranges it nearly ascends to the alpine zone. Despite the common name, the Western Terrestrial Garter Snake is often found near water. The Mountain Garter Snake and Coast Garter Snake subspecies tend to be more terrestrial, and they are often seen on the dry slopes of hills far from water. Occasionally, they will be encountered in water, but they will usually try to escape into grass and brush on the shore (except for Oregon's Willamette Valley populations, which are highly aquatic). The Wandering Garter Snake is generally found near water throughout most of the warm season, especially in the more arid climate east of the Cascade Mountains. There, it is rarely seen far from the vicinity of vegetated riparian zones. When alarmed, it usually escapes into the water, often diving to the bottom and hiding under submerged rocks and other available retreats. Their rocky, south-facing winter hibernation sites are sometimes considerable distances from the summer aquatic habitats and migrations are required. At these times large numbers may be seen crossing roads in the spring and autumn. The Western Terrestrial Garter Snake is known to often hibernate with other species of snakes, including the Western Rattlesnake. Reflecting the diversity of habitats frequented by this species, a wide variety of foods are eaten: small fish, amphibians and their larvae (salamanders, frogs, and toads), leeches, slugs, snails, earthworms, lizards, snakes, small mammals, and birds. Four to 19 young are live-born during August or early September.

FIELD NOTES: 28 May 1980. While visiting my parents in the Greenhorn Mountains of northeastern Oregon, I happened to hear the plaintive squeaking of a mouse just outside the door of their log cabin. Tracing the sound to a clump of grass by the creek,

Adult Klamath Garter Snake. Lakeview, south-central Oregon.

Adult Mountain Garter Snake; Willamette Valley variation. Flying M Ranch, North Yamhill River, northwestern Oregon.

I cautiously peered into the screening vegetation. Within arm's reach, I saw a medium-sized Western Terrestrial Garter Snake that was in the process of attempting to overpower and eat a Long-tailed Vole (*Microtus longicaudus*). Fearing that the snake might notice me and flee from its meal, I slowly sank to a kneeling position and watched the battle. Although there are no true constrictors in the *Thamnophis* genus, this individual was doing a credible job of impersonating one. It had two coils wrapped around the small rodent, which was struggling admirably. Although unable to lethally arrest the vole's breathing, the snake's entwined body was efficiently keeping its prey from escaping. Additionally, I noticed that it was repeatedly biting and chewing on the mouse. The Western Terrestrial Garter Snake has enlarged rear teeth and modified salivary glands that secrete a mild toxin (harmless to humans) that aids in subduing fighting, biting prey. Within a few minutes the vole was successfully swallowed. I walked back to the cabin feeling fortunate to have witnessed two effective tactics used by this species of snake to provide itself with a meal.

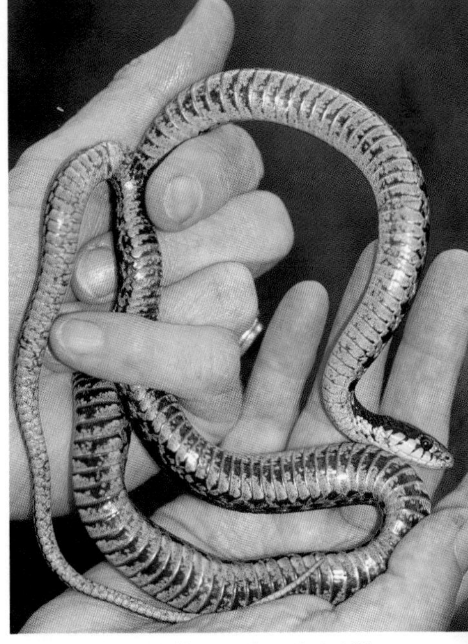

Underside of an adult Mountain Garter Snake; Willamette Valley variation. Flying M Ranch, North Yamhill River, northwestern Oregon.

Pacific Coast Aquatic Garter Snake

Thamnophis atratus

Adult; gray variation of the spotted morph. Applegate River drainage, Siskiyou Mountains, southwestern Oregon.

IDENTIFICATION: Adults are 18–33 in (46–84 cm) long. **There are usually eight upper labials (lip scales) on each side of the head. The internasal scales on the top of the snout are longer than wide and pointed at the front end.** As with all species of garter snakes, the dorsal scales on the body are strongly keeled. The head is somewhat narrow and has a slightly pointed snout. There are two distinct color/pattern variations found within the region covered by this guide: a spotted morph and a striped morph. The spotted form occurs throughout the entire range of this species in our area. It has a pale gray to olive-gray or olive-brown ground color, which is often flecked with light green. There also are two rows of squarish black spots along each side of the body that are arranged in a checkered pattern. The dorsal stripe is dull yellow, narrow, and faint, often visible only on the neck. There may be a faint yellowish stripe along each lower side of the body, as well, but these markings are frequently lacking. When seen swimming through water or sunning on a distant rock in a stream, the first impression is of a gray, dark-checkered snake that has no stripes. The striped form occurs in northern California, often coexisting with the spotted morph in the same population. It has a gray ground color that is so dark that the black-spotted pattern is not readily discernible. The lengthwise yellow stripes are well-defined (the side stripes are often brighter and wider than the dorsal stripe), which gives the appearance of a black snake with three yellow stripes. **Ventrally, both morphs lack dark markings, grading from light gray on the belly to orange or pinkish purple on the underside of the tail (sometimes pale or absent).**

VARIATION: Of the two subspecies currently recognized, only one occurs in the Northwest: the OREGON GARTER SNAKE (*T. a. hydrophilus*). The Pacific Coast Aquatic

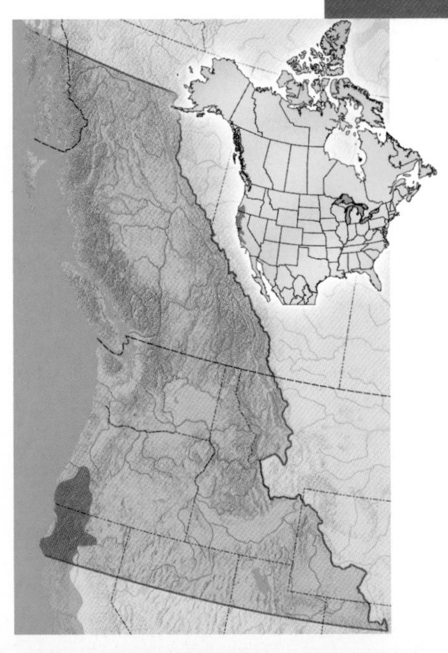

Garter Snake was formerly considered a subspecies of the Sierra Garter Snake (p. 232). These two species are closely related, and they appear to hybridize (interbreed) in the Pit River drainage of Shasta County in northern California. Many individuals from that area have characteristics of both species.

SIMILAR SPECIES: The five species of garter snakes native to the Northwest have similar patterns of lengthwise stripes and can be difficult to differentiate. If in doubt, use the Quick Key to Garter Snakes (pp. 68–69). The Striped Whipsnake (p. 166), California Whipsnake (p. 170), and Western Patch-nosed Snake (p. 178) also have striped patterns, but they are easily identified by their large, smooth dorsal scales. All garter snakes have strongly keeled scales.

DISTRIBUTION: The Pacific Coast Aquatic Garter Snake is found in the Klamath Mountains region at elevations from

Adult; olive-brown variation of the spotted morph. Rough and Ready Creek, Siskiyou Mountains, southwestern Oregon.

Adult; striped morph. Near Redding, northern California.

Adult. Applegate River drainage, Siskiyou Mountains, southwestern Oregon.

sea level to around 6,300 ft (1,920 m). It occurs throughout northern California west of the Cascade Mountains, except in a narrow strip along the coast from the Eureka area northward to the vicinity of the Klamath River. In Oregon, this reptile is confined to the river drainages of the southwestern portion of the state, as far north as the Umpqua Valley in Douglas County.

HABITAT AND BEHAVIOR: No other snake in the Northwest is so closely tied to water as is this species. It is rarely encountered far from a clear-flowing, rocky watercourse, where it can usually be found along open, sunny sections. Rocks on the shore or in mid-stream are used for basking, and it will often be seen swimming. The Pacific Coast Aquatic Garter Snake inhabits the rugged country of the rocky canyons, foothills, and interlacing valleys that typify the Klamath Mountains area. The plant associations native to its habitat are the usual riparian zone species of various sedges, willows, cottonwoods, maples, and alder. On the surrounding slopes are woodlands of mixed conifers, oaks, Pacific Madrone, Golden Chinquapin, and brushy chaparral. When escaping, this snake will invariably dive beneath the surface of the water and hide under submerged rocks. Biting is usually employed to defend itself when captured, along with the typical garter snake defense of releasing feces and a bad-smelling musk from glands at the vent. As would be expected of such an aquatic snake, fish and the adults and larvae of amphibians comprise the largest portion of the diet. It coexists in most places with the Foothill Yellow-legged Frog (*Rana*

Underside of an adult, showing orange color on the tail. Patrick Creek, Siskiyou Mountains, northern California.

Adult spotted morph basking on a rock. Applegate River, Siskiyou Mountains, southwestern Oregon.

Doug Knutsen photographing a Pacific Coast Aquatic Garter Snake in typical habitat. Applegate River drainage, Siskiyou Mountains, southwestern Oregon.

boylii) and adults and tadpoles of this amphibian are important food sources, as are larvae of the Pacific Giant Salamander (*Dicamptodon tenebrosus*). Terrestrial salamanders (plethodontids) have also been reported as being eaten. From 3 to 12 young are live-born during late August, September, or early October.

FIELD NOTES: 6 June 1999. While exploring the edges of Rough and Ready Creek in the Siskiyou Mountains of southwestern Oregon, I came upon a large pool of shallow water that was separated from the main streamcourse. The morning sunshine had warmed the water, and I immediately noticed my first reptile of the day. It was a juvenile Pacific Coast Aquatic Garter Snake that was busily slithering about in the pool while attempting to capture something in its small jaws. Carefully moving forward, I was able to get quite close, because the little snake was too distracted with trying to acquire its meal. I could now see that it was chasing minnows, but without much success. I continued my observations of its efforts, and after about 20 minutes it finally caught and ate one of the tiny fish. There is recent documentation that juveniles of this species will sometimes extend their quivering tongues onto the surface of water to mimic moving insects as a lure for young fish. I would have relished witnessing and photographing such behavior, but this particular snake did not use this ambush tactic while I observed it. Moving onward up the stream, I came to the main channel where the water was fairly deep and swift. Upon reaching the shore, a large adult of the same species of snake dove off a rock where it had been basking and disappeared into an underwater crevice. I had taken only two more steps along the stream when I heard a "kerplunck." It was a Yellow-legged Frog that had hopped into the water almost from under my feet. More than likely, the snake had been stalking the amphibian. What I had witnessed that morning was typical for this aquatic-adapted garter snake. Juveniles feed on the smaller fish and frog tadpoles in the side pools, whereas the adults are able to hunt for larger prey in the strong currents of the main sections of streams and rivers.

Adult striped morph. The Nature Conservancy's McCloud River Preserve, northern California.

The light belly is typical of the Pacific Coast Aquatic Garter Snake, but the tail lacks the usual orangish coloration of the species. The Nature Conservancy's McCloud River Preserve, northern California.

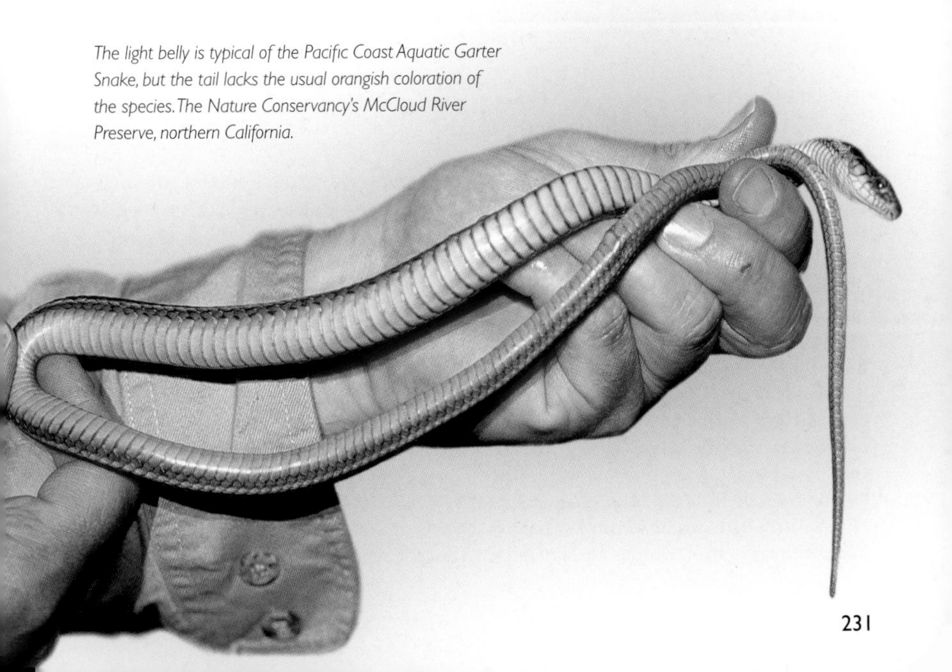

Sierra Garter Snake

Thamnophis couchii

Adult. Susan River, eastern Sierra Nevada, northern California.

OTHER NAMES: Western Aquatic Garter Snake.

IDENTIFICATION: Adults are from 18 in (46 cm) to slightly over 36 in (91 cm) in length. **There are usually eight upper labials (lip scales) on each side of the head. The internasal scales on the top of the snout are longer than wide and pointed at the front end.** As in all garter snakes, the dorsal scales on the body are strongly keeled. **The head is narrow, with a distinctively pointed snout, and the eyes are relatively large.** A narrow, dull yellow dorsal stripe usually runs down the middle of the back, but it is often rather faint and is most apparent on the neck. There may or may not be a yellowish stripe along each lower side of the body as well. The ground color between the stripes is generally olive-brown, dark grayish brown, or nearly black, with a pattern of two rows of alternating black spots. Ventrally, the belly is yellowish gray or greenish gray, with **considerable dark charcoal to black mottling, often becoming nearly solid black on the underside of the tail.**

VARIATION: Although there are some morphological differences between populations in our region and those from the southern portions of this snake's range, no subspecies are recognized. It is closely related to the Pacific Coast Aquatic Garter Snake (p. 226) and the two species appear to hybridize (interbreed) in the Pit River drainage of Shasta County. Many individuals from that area have characteristics of both species.

SIMILAR SPECIES: The five species of garter snakes native to the Northwest have similar patterns of lengthwise stripes and can

be difficult to differentiate. If in doubt, use the Quick Key to Garter Snakes (pp. 68–69). The Striped Whipsnake (p. 166), California Whipsnake (p. 170), and Western Patch-nosed Snake (p. 178) also have striped patterns, but they are easily identified by their large, smooth dorsal scales. All garter snakes have strongly keeled scales.

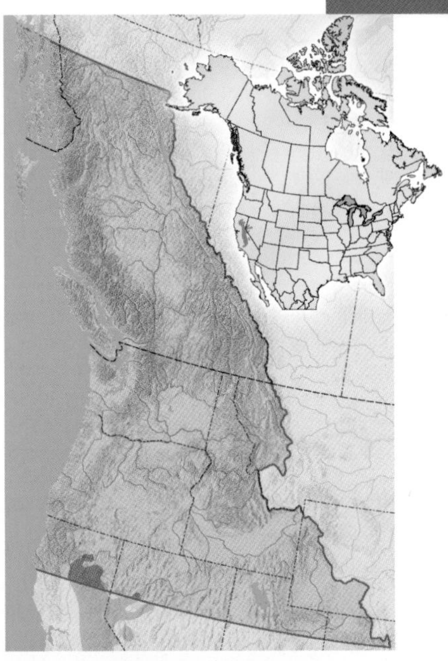

DISTRIBUTION: As its common name implies, the primary range of this species is in the Sierra Nevada of California. The Sierra Garter Snake reaches the northern limits of its distribution in the area covered by this guide. It occurs at elevations of around 600–6,000 ft (180–1,830 m), from the northern Sierra Nevada to the southern end of the Cascade Mountains in the Pit River drainage. There are also populations in a small portion of west-central Nevada along the Truckee and Carson rivers, just to the south of our area. At present, there are no records for this snake in northwestern Nevada.

Adult. Susan River, eastern Sierra Nevada, northern California.

Sierra Garter Snake

HABITAT AND BEHAVIOR: The Sierra Garter Snake is highly aquatic, and it is usually found along rocky streams and rivers and around ponds and lakes. It occurs from the mixed oak savannas and brushy chaparral woodlands of the western Sierra foothills upward in elevation to the montane coniferous forests. Along the eastern slope of this mountain range, it is found around aquatic habitats in pine, juniper, and sage-brush communities. During the morning, look for this reptile sunning on grass clumps, logs, and rocks on the shores of streams and lakes. By late morning or early afternoon, when its body temperature has risen, the Sierra Garter Snake can be seen in the side pools and quieter stretches of watercourses hunting for prey. Herpetologists who have studied this species in the field report observing these snakes crawling along the bottoms of streams searching among aquatic plants and submerged leaves for small creatures to eat. It feeds on fish and both the adult and larval forms of frogs, toads, and salamanders. Other field studies have found that larger prey items, such as trout and adult amphibians, are usually dragged up on shore to be swallowed. There is little recorded on the reproductive habits of the Sierra Garter Snake. Some recent data indicate that from 5 to 38 young are live-born during late July, August, or early September in the foothills of the Sierra, whereas at higher elevations the birthing period is late August through September.

FIELD NOTES: 9 June 1999. Today, I once again experienced the benefits of taking time to contact the right herpetological authority to gain pertinent information. Help is nearly

Underside of an adult, showing the dark belly and the lack of orange on the tail. Susan River, eastern Sierra Nevada, northern California.

Adult basking on a rock at the edge of the Susan River, eastern Sierra Nevada, northern California.

always required when dealing with the taxonomically confusing garter snakes. In this instance, I needed to photograph a Sierra Garter Snake for this field guide, a species with which I had no experience. The year before I had failed completely when I attempted to find it along the McCloud River in Shasta County of northern California. I had captured two snakes, but after examining them I realized they were actually Pacific Coast Aquatic Garter Snakes, a closely related species. After two unproductive attempts southward in the Sierra, I had to give up. The following spring, just before leaving on this outing, I phoned Dr. Steve Arnold at Oregon State University, who I knew was heading up an extensive taxonomic study of western garter snakes. When I explained my need, he said, "Oh sure, I know a place where you can always find *couchii*." Simple as that, and he proceeded to give me directions to a site on the Susan River in the eastern Sierra. So, with the able assistance of Doug Calvin, I easily found Steve's garter snake hot spot, and within 20 minutes I captured two specimens just where he said they would be. Both were typical examples of the Sierra Garter Snake, and I got my photographs. I should have phoned him the year before and saved myself a lot of time and trouble.

Ground Snake

Sonora semiannulata

Adult; orange-and-black, crossbanded morph. Near Adrian, Owyhee River Canyon, southeastern Oregon.

IDENTIFICATION: Adults are small, averaging 8–12 in (20–30 cm) long, with occasional individuals reaching 18 in (46 cm). **There is considerable variability in the dorsal coloring and pattern of this snake.** In many populations, the most common morph is one with alternating black and reddish-orange bands that cross the back (but do not extend across the belly). Some individuals may have a uniform light tan, dark tan, or olive-brown dorsal coloration, while others may have this same ground color in combination with a narrow to wide lengthwise stripe of orange down the middle of the back. Less commonly, a variation with alternating black and white or black and gray crossbands is found. No matter what pattern and coloring, **all Ground Snakes have a single distinguishing characteristic: a dark spot at the center of each dorsal scale that gives them a dappled appearance (most apparent on the sides of the body). The belly is usually an unmarked white or pale yellow.** Ground Snakes are well equipped for a burrowing life with **smooth, glossy scales and a narrow, short head that is barely wider than the neck.**

VARIATION: Because of this snake's great diversity of patterns and coloring across its broad range in the American West, two species, each with several subspecies, had been named in the past. Later, it was discovered that most or all of these variations could be found in a single population. There is now considered to be only one species with no subspecific divisions.

SIMILAR SPECIES: The Long-nosed Snake (p. 206) has a dorsal pattern that is similar to the crossbanded morphs of the Ground Snake, and both species often share the same habitat. The Long-nosed Snake can be differentiated by its mostly single, undivided row of caudal scales on the underside of the tail. Ground Snake morphs that have a lengthwise narrow stripe of orange down the

middle of the back could possibly be confused with a garter snake (pp. 210–235), but all species of garter snakes have strongly keeled scales.

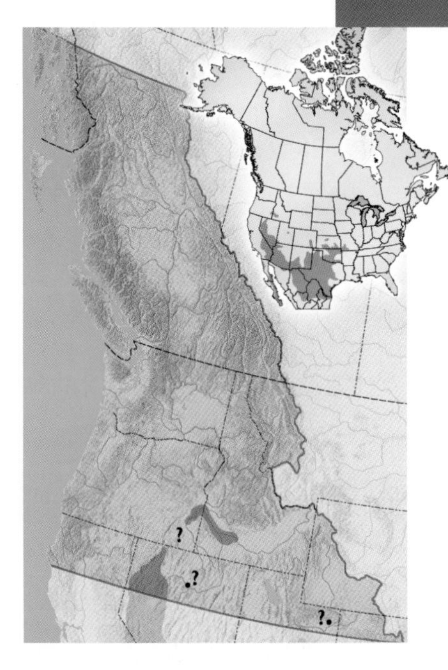

DISTRIBUTION: The Ground Snake reaches the northern limits of its distribution in our region. In the western Great Basin of Nevada it has been recorded as far north as near Denio, Humboldt County. In addition, this species is reported to occur in the vicinity of Battle Mountain, Lander County, Nevada. There are also populations in the Snake River drainage of southwestern Idaho and along the Owyhee River in southeastern Oregon. These populations appear to be disjunct from the primary range in Nevada, but future field surveys may fill in the gaps between the two areas. It is probable that this snake will someday be found in the Honey Lake Basin of Lassen County in northeastern California. There are two 1928 records for this species in northeastern Utah, near Vernal, Uintah County, but updated verification is needed. Elevation records for the Northwest range from 2,300 ft (700 m) in Idaho's Snake River Valley, to around 4,500 ft (1,370 m) in northern Nevada.

HABITAT AND BEHAVIOR: These little burrowing snakes inhabit shrubby desert regions, preferring areas with loose, sandy soils. The slopes along the edges of washes where there are scattered rocks seem to be especially favored. Typical plants in the habitat are Big Greasewood, Shadscale, Four-winged Saltbush, and Spiny Hopsage.

Juvenile; white-and-black, crossbanded morph. Near Adrian, Owyhee River Canyon, southeastern Oregon.

Ground Snake

Adult; orange-striped morph. Near Adrian, Owyhee River Canyon, southeastern Oregon.

Adult; uniform dark tan morph. Near Boise, lower Snake River Valley, southwestern Idaho.

Adult; uniform tan morph. Near Boise, lower Snake River Valley, southwestern Idaho.

Ground Snakes lead a mostly subterranean life and are active on the surface only during the late evening and at night. These secretive habits make them one of the most difficult-to-find reptiles in the Northwest. They can sometimes be found under rocks during the springtime or after a summer rainstorm when there is some surface moisture. During dry periods they are rarely encountered, except for the occasional individual seen crossing a road at night. The reddish-orange and black crossbanded morphs are sometimes mistakenly thought to be a venomous coral snake, but Ground Snakes are harmless and never attempt to bite when handled. Insects and their larvae, spiders, centipedes, and scorpions have been recorded as being eaten. The rear teeth are equipped with shallow grooves and there is some indication that the Ground Snake may be mildly venomous to aid it in subduing small prey. Little is known about the breeding habits of this egg-laying species. Present information indicates that up to six eggs are laid, probably during July and August in the Northwest.

FIELD NOTES: 21 June 1984. While conducting a herpetological inventory of southeastern Oregon's Owyhee River drainage for the Oregon Department of Fish and Wildlife, a drenching thunderstorm brought me good fortune. I was attempting to find new localities for the Ground Snake. On the prior morning I had been lucky enough to capture one that was under a large stone at a remote site further up the river. It was hotter today and I doubted that I'd have much success looking for this species under surface rocks, so I had planned to switch to roadhunting at night. This method had proved successful during August of 1978, when Doug Knutsen had discovered two dead Ground Snakes on a paved road along the Owyhee. One was an unmarked tan morph. Prior to that, only the black-and-orange crossbanded variation had been found in Oregon. However, I had to reconsider my night-driving plan when a spectacular noon-time downpour today had quickly changed conditions. It had become cooler and relatively humid for this arid country, so I decided to try rock flipping again. Driving to the exact spot where Doug had found the two *Sonora* six years earlier, I parked and began climbing up a steep slope directly above the road. Upon turning the first rock, I became optimistic when I saw the moist, sandy soil that was revealed. I immediately began finding Ground Snakes, and within an hour's time I collected five adults and two juveniles. What really excited me, though, was that along with the usual orange-and-black crossbanded individuals, there was one with a wide, orange, lengthwise stripe down its back, and another that had black and white crossbands. Counting Doug's 1978 tan morph, there was now confirmation that Oregon has all four of the color/pattern morphs known to occur in the Northwest. After taking photos, it was gratifying to release the colorful assemblage and watch them disappear beneath rocks.

Underside of an adult. Near Boise, lower Snake River Valley, southwestern Idaho.

Habitat of the Ground Snake along a remote stretch of the Owyhee River Canyon in southeastern Oregon. This area was dubbed "Ground Snake Gulch" by the author and friends after several successful field trips there in the late 1960s.

Night Snake
Hypsiglena torquata

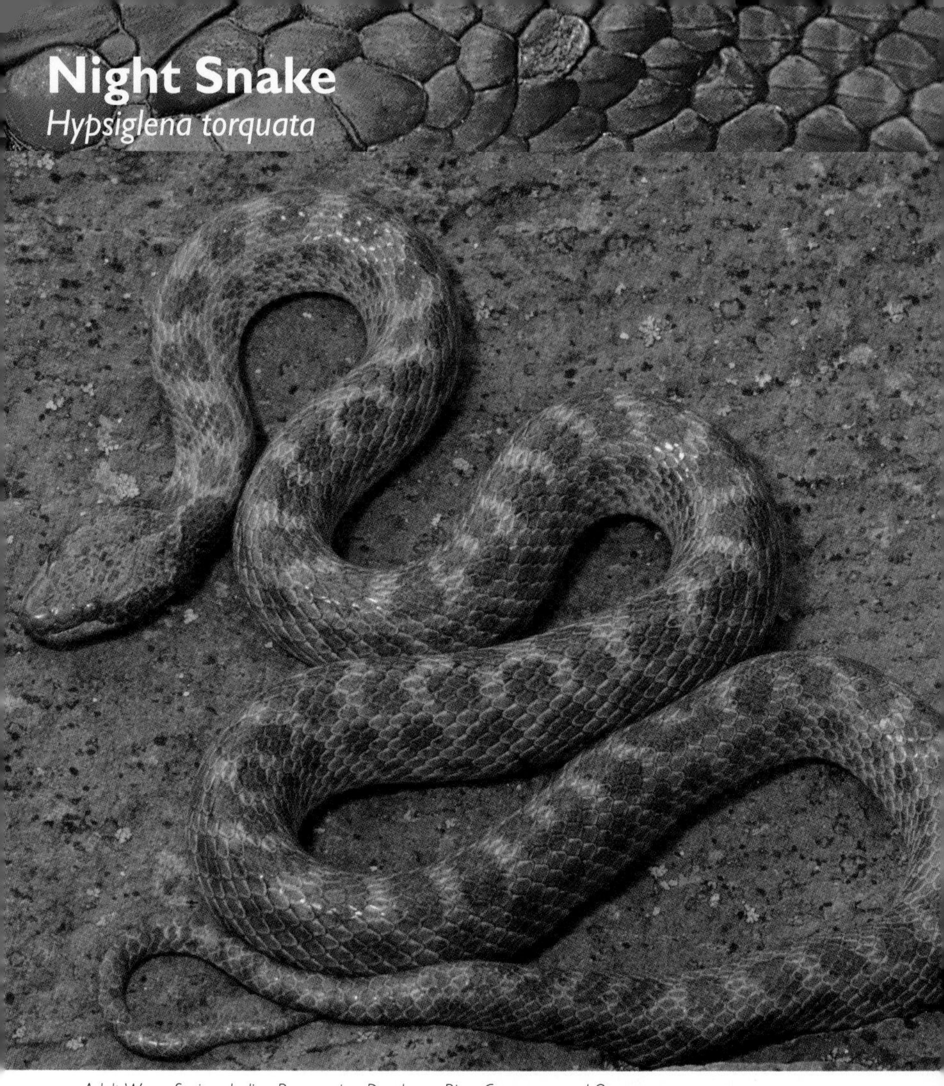

Adult. Warm Springs Indian Reservation, Deschutes River Canyon, central Oregon.

IDENTIFICATION: Most adults are small, usually averaging 12–18 in (30–46 cm) long. Occasional individuals may slightly exceed 24 in (61 cm). Dorsally, the ground color is tan, gray, light brown, or pinkish beige. There is a pattern of dark brown, rather squarish blotches along the middle of the back, with a corresponding row of smaller blotches along the sides of the body. **A distinctive, dark, collar-like marking covers the neck, which may be divided into two or three sections.** There is a dark facial mask stripe on each side of the head. The head is relatively flat and wide, and **the eyes have vertical pupils.** Ventrally, the belly is glossy white or yellowish and lacks any darker markings. The dorsal scales are smooth.

VARIATION: There are differences in the neck markings, head scales, and other characteristics throughout the range of the Night Snake, and several subspecies have been described. Presently, many herpetologists question the validity of these divisions, and further studies are needed. Three subspecies are considered to occur in our region: the MESA VERDE NIGHT SNAKE (*H. t. loreala*) in northeastern Utah; the CALIFORNIA NIGHT SNAKE (*H. t. nuchalata*) in the Sacramento Valley of northern California; and the DESERT

NIGHT SNAKE (*H. t. deserticola*) in the remainder of this species' range in the Northwest.

SIMILAR SPECIES: The Gopher Snake (p. 182), juvenile Racer (p. 162), and Corn Snake (p. 186) all have blotched dorsal patterns, but they differ in having eyes with round, not vertical, pupils. The Western Rattlesnake (p. 244) has vertical pupils and a blotched pattern, but it can be differentiated by its strongly keeled dorsal scales and the rattle on the end of the tail.

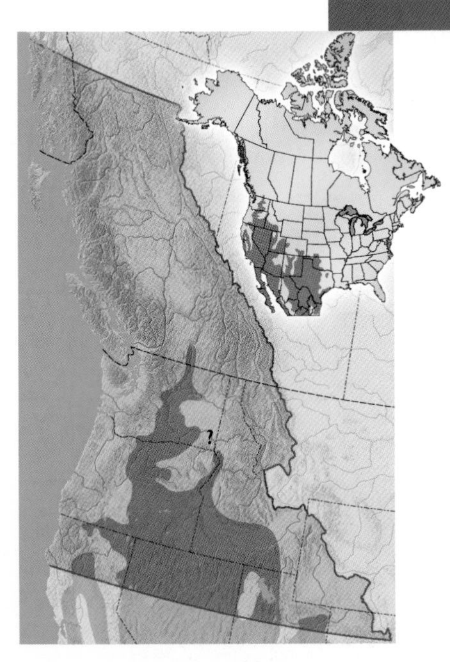

DISTRIBUTION: In the Northwest, the Night Snake ranges throughout a great portion of the interior plateau east of the Cascade Mountains. It occurs where there is suitable habitat in northern Nevada, northeastern California, northern Utah, southern Idaho, central and eastern Oregon, southeastern and central Washington, and extreme south-central British Columbia (in the southern Okanagan Valley). This snake is also found west of the Cascade-Sierra Mountains in the Sacramento Valley of northern California. Additionally, it has been recorded in Siskiyou County in extreme northern California, where it penetrates west of the Cascades along the Klamath River Canyon and into the adjacent Shasta Valley. The Night Snake occurs in our region at elevations from just above sea level in the Columbia River Gorge of Washington and Oregon, to nearly 6,000 ft (1,830 m) in the Uinta Basin of northeastern Utah. Although there are no confirming records, it is probable that this species is also found in northwestern Colorado's Green and Yampa river drainages.

HABITAT AND BEHAVIOR: This is a snake of the arid and semi-arid sections of the Northwest. It ranges throughout all but

Adult. Owyhee River Canyon, southeastern Oregon.

the higher, more forested parts of the Great Basin, occurring in desert shrublands where such typical plants as Big Greasewood, Shadscale, Four-winged Saltbush, and Spiny Hopsage predominate. Canyonlands and rimrock country in sagebrush-juniper communities are also inhabited, sometimes at the fringes of open pinewoods. In northern California and at the eastern end of the Columbia River Gorge, this species is found in dry oak savannas, mixed pine-oak woodlands, and chaparral. The primary habitat of the Night Snake is rocky areas, but it is sometimes encountered in sandy desert basins, where it seeks refuge in rodent burrows. As the common name indicates, this reptile is active only at night, and although it is actually quite widespread, it is rarely seen. It can sometimes be found by looking under rocks, but usually the more successful method is to slowly drive roads after dark. Individuals will often cross blacktop on nights that are unusually cool for reptile activity. When threatened, the Night Snake will sometimes coil into a tight circle, flatten its head, and strike outward without trying to bite. Foods consist mainly of lizards (especially Common Side-blotched Lizards), but lizard eggs, small snakes, salamanders, frogs, and toads have been recorded as being eaten as well. Venom is secreted with the saliva of this snake, which flows down grooves on its enlarged rear teeth when it bites prey. Two authors have both reported observations of Common Side-blotched Lizards succumbing

Adult, showing the vertical pupil of the eye. Malheur National Wildlife Refuge, southeastern Oregon.

within 10 minutes after being captured in the jaws of Night Snakes. However, the venom injected is of such a minimal amount that it poses no threat to humans. The meager information available on the Night Snake's breeding habits in the Northwest indicate that three to nine eggs are laid in late June or early July and probably hatch during late August or early September.

Underside of a small adult. Owyhee River Canyon, southeastern Oregon.

FIELD NOTES: 20 June 1984. After two hours of turning rocks along the edge of a dry wash, I stopped and straightened to give my aching back a rest. Wiping the sweat from my eyes, I admired the rugged scenery in this remote section of Oregon's Owyhee River Canyon. Late June can be quite hot in the desert and, consequently, is usually not the most profitable time of year to search under rocks for snakes. The late morning sun was becoming more intense, so I decided to continue hunting for a few more minutes and then quit for lunch. I had turned only two or three stones when I found a 20-in (51-cm) Night Snake, and then a couple of rocks later, a 10-in (25-cm) Ground Snake. Pleased with my luck, I walked back to my VW camper van, folded down the mini-table, and sat in the welcome interior shade. I decided to place the two snakes in a glass jar so that I could observe them while I ate a sandwich. About 10 minutes later, I noticed I was not the only one having a meal. The Night Snake had grasped the little Ground Snake in its jaws. Fascinated, I watched as the struggle ensued, the Ground Snake writhing about in a valiant attempt to free itself. Within 10 minutes I noticed that its struggles were weakening. I doubt, though, that this was due entirely to the effects of the Night Snake's venom. The Ground Snake had been grasped at its neck by the Night Snake and was probably being strangled. Shortly, the small orange-and-black, banded snake died and was slowly swallowed. Both species often exist in the same habitat, and it is probable that Ground Snakes are frequently eaten by Night Snakes. Our shared lunchtime was over, so I took the bulging snake back to the capture site and watched it slither into a hiding place where its meal could be digested in peace.

Western Rattlesnake
Crotalus viridis

Adult Northern Pacific Rattlesnake; olive-tan variation. Trinity River drainage, northwestern California.

IDENTIFICATION: Adults average 15–36 in (38–91 cm) long, occasionally reaching 48 in (1.2 m). Rarely, 60-in (1.5-m) individuals are encountered. This snake is the only venomous reptile native to the Northwest that is dangerous to humans. It is easily identified by the **short, thick tail with distinctive rattles**, which allow it to produce a startling "buzz." **Recently born juveniles are equipped with only a silent nubbin on the tip of the tail**, called a "button." Also diagnostic are the **wide, triangular head, vertical pupils, narrow neck, stout body, and strongly keeled scales.** The Western Rattlesnake typically has a basic dorsal pattern of large, brown blotches down the middle of the back, with a corresponding row of smaller blotches along each side of the body. The overall ground color of the snake's body may be tan, golden tan, brown, reddish brown, gray, yellowish, or olive green. The dorsal and side blotches join toward the rear of the snake and form crossbands, culminating as encircling rings on the tail (over two-thirds of the body may be crossbanded on many individuals from Oregon's Willamette Valley). Juveniles have a brighter dorsal pattern than adults.

VARIATION: Although it is the Northwest's only indigenous rattlesnake, the extreme variability in the Western Rattlesnake's appearance has led to the erroneous belief that several species are found in the region, such as "timber rattlers," "diamondbacks," and "sidewinders." Blending remarkably well with the surrounding habitat, Western Rattlesnakes from wooded areas are generally darker and have larger blotches, whereas those from the open deserts and plains are lighter, with smaller blotches. Dark, melanistic rattlesnakes are found in some areas with extensive black lava fields. Eight subspecies are currently recognized, with four occurring in the Northwest. The NORTHERN PACIFIC RATTLESNAKE (*C. v. oreganus*) has large, closely spaced, squarish, light-edged dorsal blotches against a ground color of brown, gray, or greenish olive, and black and white tail rings. It occurs west of the Cascades

in California and Oregon, and to the east of that mountain range in the Columbia Plateau region as far north as south-central British Columbia. The GREAT BASIN RATTLESNAKE (*C. v. lutosus*) has relatively wide-spaced, oval, dark-edged dorsal blotches against a ground color of buff, tan, or golden tan, and no white tail rings. It occurs from southeastern Oregon and southern Idaho southward through northeastern California, Nevada, and western Utah. The PRAIRIE RATTLESNAKE (*C. v. viridis*) is similar in appearance to the Great Basin form, but its dorsal blotches are larger and light edged, and the ground color is often greenish or yellowish. It also has no white tail rings, and it enters our area only in western Montana, northwestern Colorado, and the upper Salmon River drainage in central Idaho. The MIDGET FADED RATTLESNAKE (*C. v. concolor*) has very pale oval blotches that are sometimes barely discernible against a slightly lighter ground color of tan, yellowish tan, or pinkish tan. It has no white tail rings, and it rarely exceeds 26 in (66 cm) in length. It occurs in the Green River drainage of northeastern Utah, extreme northwestern Colorado, and southwestern Wyoming.

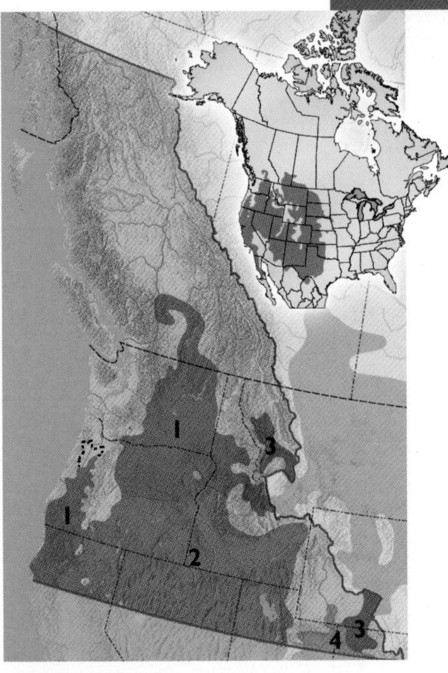

SUBSPECIES

- **1** Northern Pacific Rattlesnake *C. v. oreganus*
- Former range (dashed outline)
- **2** Great Basin Rattlesnake *C. v. lutosus*
- **3** Prairie Rattlesnake *C. v. viridis*
- **4** Midget Faded Rattlesnake *C. v. concolor*
- Species range outside the Pacific Northwest

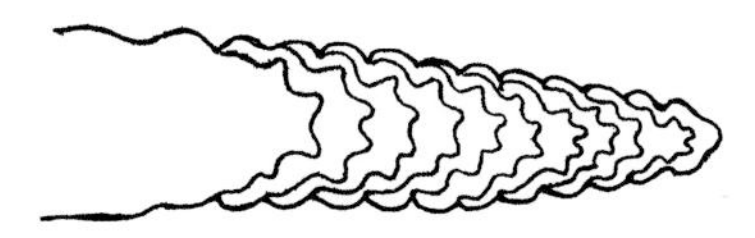

Cutaway view of rattle showing interlocking segments

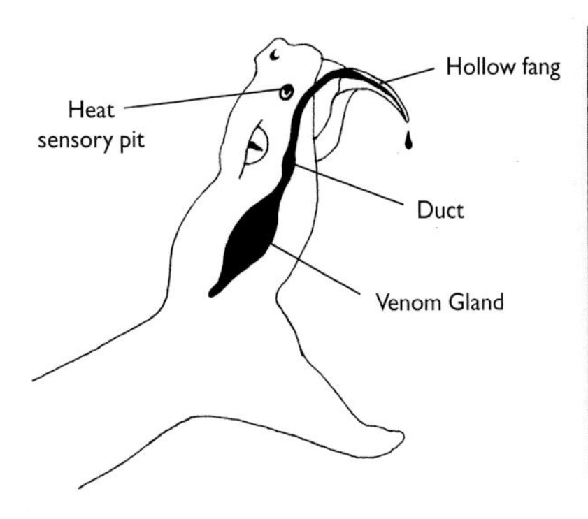

Adult Northern Pacific Rattlesnake; brown variation. Deschutes River Canyon, near Madras, central Oregon.

Western Rattlesnake

Adult Great Basin Rattlesnake; light tan variation. Pyramid Lake, northwestern Nevada.

SIMILAR SPECIES: The Gopher Snake (p. 182), Corn Snake (p. 186), juvenile Racer (p. 162), and Night Snake (p. 240) all have blotched patterns, but the Western Rattlesnake's unique rattle or button on the end of the tail quickly confirms its identity.

DISTRIBUTION: Except for the higher, densely forested mountains above approximately 5,000–7,000 ft (1,520–2,130 m), this snake occurs throughout most of the drylands east of the Cascade Mountains. It also inhabits the interior valleys of northern California and western Oregon, but it is entirely absent west of the Cascades in Washington and British Columbia. The area enclosed by dashes on the distribution map indicates the borders of the rattlesnake's former range in the northern Willamette Valley of Oregon.

Adult Great Basin Rattlesnake; dark tan variation. Surprise Valley, northeastern California.

HABITAT AND BEHAVIOR: The wide-ranging Western Rattlesnake lives in a variety of ecosystems in the Northwest, favoring dry, rocky, brushy places, and open savannas. The Northern Pacific subspecies inhabits both the oak woodlands west of the Cascades and the juniper and pine associations of the high intermountain plateaus. In the sagebrush rangelands and deserts, the Great Basin subspecies is found, while the Prairie Rattlesnake is native to the grassy plains at the eastern border of our region. The Midget Faded Rattlesnake occurs in sandstone canyonlands of the Green River area, amid stunted forests of Utah Juniper

and Two-needle Pinyon. Western Rattlesnakes use rocky southern exposures as hibernation dens, wherein large numbers often congregate in deep crevices. In areas of grassy plains with no rocks, the Prairie subspecies uses the burrows of prairie dogs as hibernacula. Small mammals (mice, rats, squirrels, and rabbits) are the primary food sources, along with occasional birds, lizards, and amphibians. During the heat of summer, rattlesnakes are primarily active in the morning and evening, or after dark on warm nights. From 1 to 25 young are live-born during late August, September, or early October.

Adult Prairie Rattlesnake. National Bison Range, western Montana.

FIELD NOTES: 3 October 1992. It was a beautiful, sunny autumn afternoon in the Willamette Valley when I made a solo trip up the steep southern slope of a rocky butte in Marion County. I have been visiting rattlesnake dens there since I was a teenager, but several years had passed since my friends and I had seen rattlers on the hill, and we feared they may now be a thing of the past. These unique reptiles were once a widespread, integral component of the Willamette environment. At present, though, they are primarily restricted to the southern portions of this heavily populated valley. Originally found from the McMinnville and Molalla areas southward, the rattlesnake now appears to have been extirpated in Benton, Polk, Yamhill, and Clackamas counties. This

Adult Midget Faded Rattlesnake. Red Fleet State Park, Uinta Basin, northeastern Utah.

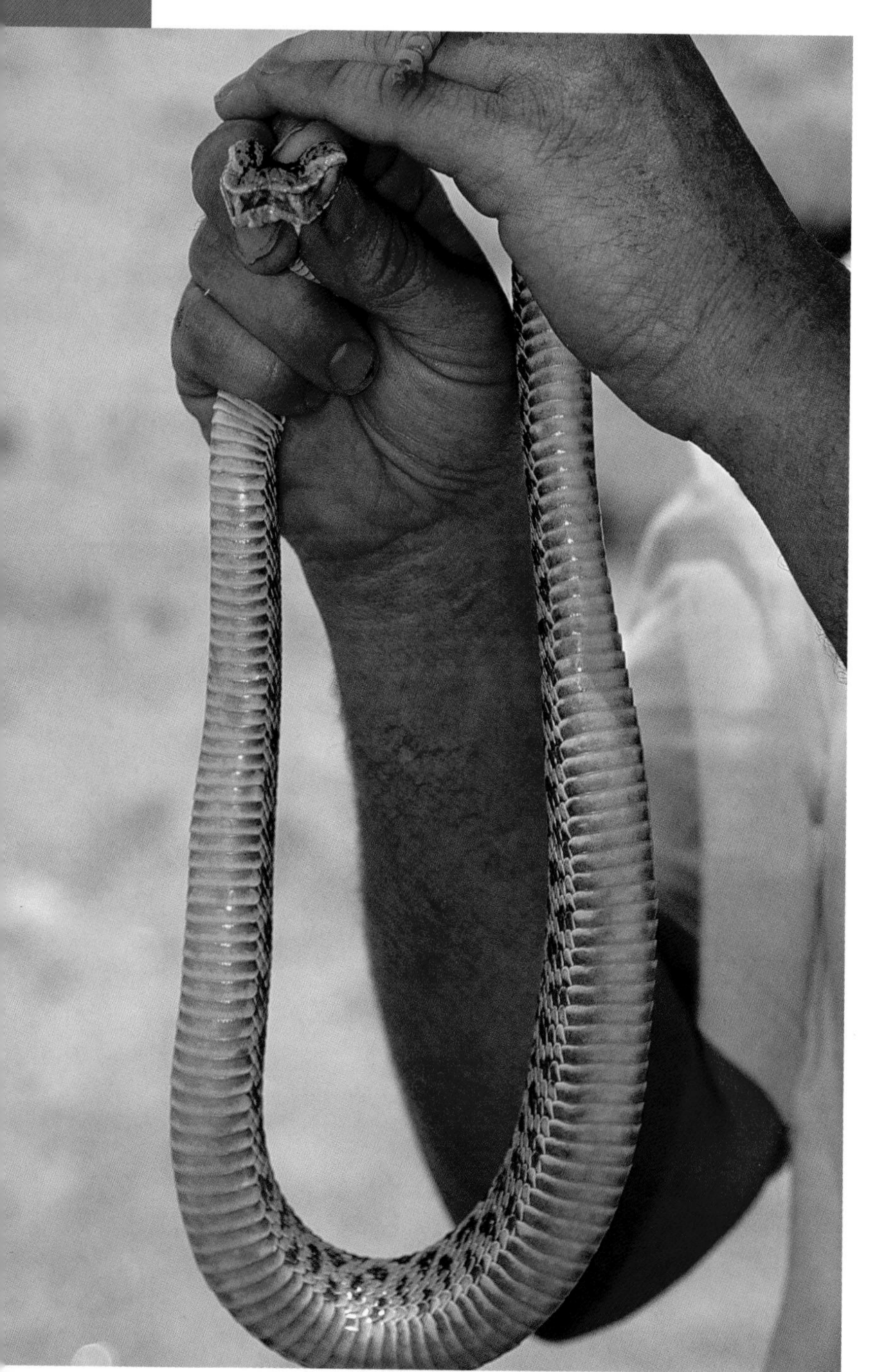

Underside of an adult Great Basin Rattlesnake. Pueblo Mountains, southeastern Oregon.

Adult Northern Pacific Rattlesnake basking in early spring sunshine at its denning site. Near Sisters, central Oregon.

oak-covered promontory is their last bastion this far north in western Oregon. My hope is that a few are still managing to survive on a more inaccessible part of the butte. As the day waned, I clambered down a nearly vertical rock face to a jumble of stones at its base and discovered two recently shed *Crotalus* skins in a crevice. With renewed enthusiasm, I continued farther along the ledges and suddenly saw it: a juvenile rattlesnake basking in the late-day sun. It was my birthday, and I can't imagine a better gift than this discovery. After photographing the small snake, I sat on a rock for a few minutes and contentedly shared its view of golden maple leaves against a blue sky, with snowy Mt. Jefferson in the eastern distance. Retracing my steps down the slope in the crisp twilight, I hoped that the Western Rattlesnake will remain undisturbed and survive in this secluded spot.

The Marion County juvenile Northern Pacific Rattlesnake. Willamette Valley, northwestern Oregon.

Adult. Captive, Brad's World of Reptiles, Corvallis, Oregon.

SNAPPING TURTLE

(Chelydra serpentina)

Native to the region of North America east of the Continental Divide, this primitive-looking turtle has been introduced into aquatic habitats in several parts of the Northwest. It has been found in the Puget Sound area of Washington and in Oregon in the Willamette Valley, near Roseburg, and on the coast at Coos Bay. There is one observation of a Snapping Turtle laying eggs in a Seattle lake, and other reproducing Northwestern populations may have become established.

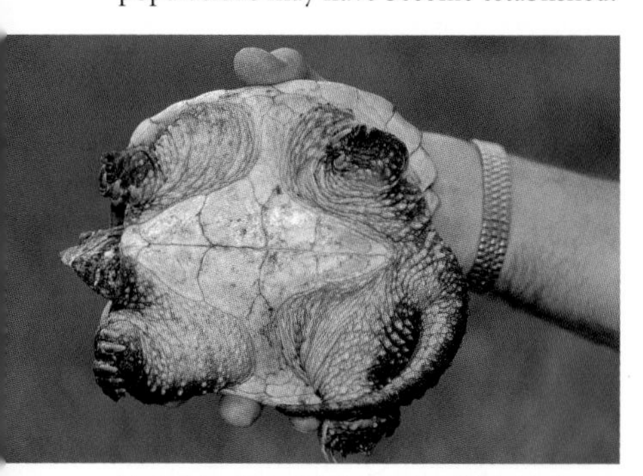

Underside of an adult. Captive, Brad's World of Reptiles, Corvallis, Oregon.

Because females can produce well over 50 eggs in a single clutch (the record is 109 eggs), populations can grow quickly. This species is large, reaching 18 in (46 cm) in shell length and weighing up to 75 lb (34 kg). The Snapping Turtle can be identified by its **massive head and legs, and a long tail that is nearly two-thirds the length of the shell. The drab-brown carapace often has protruding, knobby keels, and the plastron is extremely small and narrow.** The mud-colored shell is usually covered with a growth of algae, so it blends well with the surrounding environment. The Snapping Turtle is also rather secretive in its habits—it rarely basks on logs or banks—so it can occur in wetlands adjacent to human habitation without being detected. The Snapping Turtle favors standing water with an abundance of aquatic plants and a mud bottom, but it is sometimes found in slow-moving sections of rivers and streams. A wide variety of foods are eaten, including fish, amphibians, crayfish, snakes, insects, small mammals, birds (particularly young waterfowl), carrion, and water plants. This reptile will readily bite, and large individuals can inflict serious wounds.

SLIDER

(Trachemys scripta)

This species occurs throughout much of eastern North America and northern South America. It is the most commonly sold turtle in pet shops and has been accidentally or intentionally released in many areas outside its natural distribution. Sliders have been observed at a number of locations throughout the Northwest, and several established breeding populations have been documented. Adults are 4–14 in (10–36 cm) in shell length. The Slider is similar in appearance to our native Painted Turtle (p. 76)—both species have yellow stripes on the neck and legs—but the Slider can be differentiated by a **total lack of red on the plastron, instead having a yellow coloration with a series of large, dark**

Adult. Captive, Brad's World of Reptiles, Corvallis, Oregon.

Plastron of an adult. Captive, Brad's World of Reptiles, Corvallis, Oregon.

blotches. Another distinctive characteristic often present on many Sliders is a **bright patch of red (sometimes yellow) on the side of the head, just behind the eye.** The carapace is dark olive to dark brown, with an intricate pattern of black and yellow markings. Juveniles have bright-green carapaces with yellow stripes. Sliders prefer ponds, lakes, sloughs, and the quiet stretches of streams and rivers where large numbers can often be seen sunning on logs. They typically occur in areas of water that have muddy bottoms and plenty of aquatic vegetation. Amphibian larvae, fish, crayfish, snails, and insects are eaten, but plant material usually predominates in the diet of adults. Two to 25 eggs are deposited during spring or summer, the timing depending upon the geographic region. Up to three clutches of eggs may be laid a year in southern climates.

PLATEAU STRIPED WHIPTAIL

(*Cnemidophorus velox*)

A lizard of the "Four Corners" area, this whiptail naturally inhabits southern Utah, southwestern Colorado, and the northern portions of Arizona and New Mexico. There is a thriving population at Cove Palisades State Park in Jefferson County of central Oregon that is thought to have been introduced during the 1960s. It has become well established, and recent studies indicate that the species is expanding in the immediate vicinity. Only one individual was needed to initiate this population, because this lizard is an all-female species that reproduces by parthenogenesis. The Plateau Striped Whiptail has a **more vivid, even-edged pattern of lengthwise yellow stripes along its back** than our similar native Western Whiptail (p. 134), which does not have an overlapping range with the Plateau Striped Whiptail in Oregon. It also differs in having **a plain white belly**, whereas the Western Whiptail has a gray or bluish-gray belly with dark spots. Juveniles of both species have a blue tail. Adult Plateau Striped Whiptails

Young adult and juvenile. Cove Palisades State Park, Deschutes River Canyon, central Oregon.

may reach a snout-vent length of nearly 4 in (10 cm) and a total length of slightly over 10 in (25 cm). In Cove Palisades State Park, at an elevation of 2,000 ft (610 m), this active reptile is fairly common on dry, rocky slopes that have open spaces with loose, sandy soil and a mixed growth of Western Juniper, Big Sagebrush, Gray Rabbitbrush, Bitterbrush, and various bunchgrasses. This closely approximates its native habitat in the American Southwest. Insects, such as grasshoppers, crickets, beetles, and caterpillars, along with spiders have been recorded as being eaten. In Colorado, three to five eggs are laid during late June or early July and hatch in August.

Older adult. Cove Palisades State Park, Deschutes River Canyon, central Oregon.

Northwest Plants

Listed below in alphabetical order are the common and scientific names of plants that are mentioned in the descriptions of the ecoprovince habitats and in the reptile species accounts. Where there is more than one common name in current usage or recent changes in scientific names, the alternate name is given in parentheses.

Bailey's Greasewood (Little Greasewood), *Sarcobatus baileyi*

Big Greasewood (Black Greasewood), *Sarcobatus vermiculatus*

Big Sagebrush, *Artemisia tridentata*

Bigleaf Maple, *Acer macrophyllum*

Bitterbrush, *Purshia tridentata*

Black Cottonwood, *Populus trichocarpa*

Black Spruce, *Picea mariana*

Blue Oak, *Quercus douglasii*

Bluebunch Wheatgrass, *Pseudoroegneria spicatum* (*Agropyron spicatum*)

Buckbrush, *Ceanothus cuneatus*

Bud Sagebrush, *Artemisia spinescens*

California Black Oak, *Quercus kelloggii*

California Pitcher Plant, *Darlingtonia californica*

California Sycamore, *Platanus racemosa*

Canyon Live Oak, *Quercus chrysolepis*

Canyon Maple, *Acer grandidentatum*

Deerbrush (Wild Lilac), *Ceanothus integerrimus*

Douglas-fir, *Pseudotsuga menziesii*

Engelmann Spruce, *Picea engelmannii*

Four-winged Saltbush, *Atriplex canescens*

Gambel Oak, *Quercus gambelii*

Golden Chinquapin, *Castanopsis chrysophylla*

Grand Fir, *Abies grandis*

Gray Pine (Digger Pine), *Pinus sabiniana*

Gray Rabbitbrush (Common Rabbitbrush), *Chrysothamnus nauseosus*

Green Ephedra, *Ephedra viridis*

Idaho Fescue, *Festuca idahoensis*

Incense-cedar, *Calocedrus decurrens*

Jeffrey Pine, *Pinus jeffreyi*

Knobcone Pine, *Pinus attenuata*

Limber Pine, *Pinus flexilis*

Lodgepole Pine, *Pinus contorta*

Mountain Hemlock, *Tsuga mertensiana*

Mountain Mahogany, *Cercocarpus ledifolius*

Noble Fir, *Abies procera*

Oregon Ash, *Fraxinus latifolia*

Oregon White Oak, *Quercus garryana*

Pacific Dogwood, *Cornus nuttallii*

Pacific Madrone, *Arbutus menziesii*

Pacific Rhododendron, *Rhododendron macrophyllum*

Paper Birch, *Betula papyrifera*

Plains Prickly Pear Cactus (Many-spined Prickly Pear Cactus), *Opuntia polyacantha*

Ponderosa Pine (Yellow Pine), *Pinus ponderosa*

Quaking Aspen, *Populus tremuloides*

Red Alder, *Alnus rubra*

Red Fir, *Abies magnifica*

Redwood, *Sequoia sempervirens*

Rocky Mountain Juniper, *Juniperus scopulorum*

Scotch Broom, *Cytisus scoparius*

Shadscale, *Atriplex confertifolia*

Silver Fir, *Abies amabilis*

Sitka Spruce, *Picea sitchensis*

Smokebush, *Psorothamnus polydenius*

Spiny Hopsage, *Grayia spinosa* (*Atriplex spinosa*)

Subalpine Fir, *Abies lasiocarpa*

Sugar Pine, *Pinus lambertiana*

Tanoak, *Lithocarpus densiflorus*

Two-needle Pinyon Pine, *Pinus edulis*

Utah Juniper, *Juniperus osteosperma*

Valley Oak, *Quercus lobata*

Vine Maple, *Acer circinatum*

Western Azalea, *Rhododendron occidentale*

Western Hemlock, *Tsuga heterophylla*

Western Juniper, *Juniperus occidentalis*

Western Larch, *Larix occidentalis*

Western Mountain Maple (Rocky Mountain Maple), *Acer glabrum*

Western Poison Oak, *Toxicodendron diversilobum*

Western Redcedar, *Thuja plicata*

Western Sword Fern, *Polystichum munitum*

White Fir, *Abies concolor*

White Manzanita, *Arctostaphylos viscida*

White Spruce, *Picea glauca*

Whitebark Pine, *Pinus albicaulis*

Glossary

Adult: A sexually mature animal.

Aestivate: See *estivation.*

Anal plate: A large scale on the underside of a snake that covers the vent at the base of the tail (divided as an adjoining pair on some species).

Aquatic: Inhabiting water.

Arboreal: Inhabiting trees and/or shrubs.

Bask: For an animal to assume a position that exposes it directly to the sun to absorb warmth.

Camouflage: A pattern/coloration that allows an organism to blend with its surrounding habitat.

Carapace: The upper shell of a turtle.

Carrion: The decaying flesh of a dead animal.

Caudals: The scales on the underside of a snake's tail.

Chaparral: Dry, brushy areas, usually found on hillsides. In the Northwest, this habitat occurs throughout northern California and southwestern Oregon.

Cloaca: A chamber that fecal, urinary, and reproductive secretions empty into, opening out through the vent.

Clutch: The total collection of eggs in one female's laying.

Cold-blooded: See *ectotherm.*

Colubrid: A snake belonging to the family Colubridae.

Constriction: A method used by some species of snakes to kill their prey. Two or more loops of the snake's body are tightly wrapped around an animal and the victim's chest cavity is compressed sufficiently to deflate the lungs and prevent the heart from beating, leading quickly to death.

Continental Divide: In North America, the crest along the Rocky Mountain chain that separates the river drainage systems that flow westward to the Pacific Ocean from those that flow eastward to the Atlantic Ocean.

Crepuscular: Active at dusk and/ or dawn.

Denning site: See *hibernaculum.*

Dewlap: A flap of skin on the neck of some lizards that can be extended. Usually more developed on males.

Disjunct population: A group of animals that occur in a location widely separated from the primary range of the same species. Also see *relict population.*

Diurnal: Active during daylight.

Dorsal: The upper surfaces of an animal.

Dorsal stripe: A lengthwise stripe running down the middle of a reptile's back. Sometimes referred to as a vertebral stripe.

Ecosystem: The entire complex of interdependent plants, animals, and other components that form a community within their environment.

Ectotherm: An animal that derives all or most of its body warmth from outside sources. Also called "cold-blooded."

Endemic: Native to only one particular geographical area.

Endotherm: An animal that self-regulates its body temperature through internal metabolic processes.-Also called "warm-blooded".

Estivation: A dormant state that some animals retreat into during prolonged hot and/or dry weather.-Sometimes referred to as "summer hibernation."

Eye cap: See *spectacle.*

Fang: An elongated, sharp tooth.

Feces: Excremental waste from an animal.

Femoral pores: A row of small openings (glands) on the underside of some species of lizards' thighs that contain a wax-like substance. Usually more apparent on males.

Form: A group of organisms that differ in pattern and/or coloring from others within the same species.-Also see *subspecies* and *morph.*

Genus: A grouping of closely related species.

Granular scales: Small, rounded scales that do not overlap, having a bead-like appearance.

Gravid: Heavy with eggs or pregnant with young.

Gular fold: A fold of skin at the rear of the throat on certain species of lizards.

Habitat: The natural ecological home in which a plant or animal typically occurs.

Hatchling: A young animal that has recently hatched from an egg.

Heat-sensory pits: Small facial indentations found on all pit vipers and some boas and pythons. These pits are extremely sensitive to infrared radiation and aid the snake in making an accurate strike at warm-blooded prey, even in the dark.

Herpetology: The study of reptiles and amphibians. Derived from the Greek word *herpes,* meaning "creeping or crawling thing."

Herpetologist: A scientist who studies reptiles and amphibians.

Herpetofauna: The amphibians and reptiles native to a geographical area.

Hibernaculum: A place into which an animal retreats for winter hibernation.-Also called a "denning site."

Hibernation: A dormant state of extremely low metabolic activity into which some animals retreat during the cold winter months.

Hybrid: The resulting offspring from the interbreeding of two different species of organisms. Unless both species are somewhat closely related taxonomically (in the same genus), this often results in sterile offspring that cannot reproduce.

Intergrade: As pertaining to subspecies, this term refers to individuals from a location where the ranges of two different geographic variations of a species merge and usually have a blending of morphological characteristics of both forms. These zones are called "areas of intergradation."

Internasals: An adjacent pair of scales on top of a snake's snout.

Invertebrate: An animal without a backbone.

Jacobson's organs: See *vomeronasal organs.*

Juvenile: Any young animal that is within the span between hatching/birth and sexually mature adulthood.

Keeled scale: A scale with a lengthwise, raised ridge down the middle.

Labial scale: A lip scale.

Larva: The aquatic stage of an amphibian before it metamorphoses to the terrestrial form. (*Pl:* larvae.)

Live-bearing: See *viviparous.*

Melanistic: A condition in which there is an abundance of dark pigment (melanin) in an animal's skin and/or fur, giving it a black or nearly black appearance. The opposite of albinism.

Mid-dorsal: Pertaining to the middle of an animal's back.

Morph: A term applied to a distinctly differing individual organism that is of a "polymorphic" species. Some species have two or more consistent color/pattern variations that occur within single populations, often appearing among the same hatching or litter of young. For examples of polymorphism, see the species accounts for the Ground Snake (p. 236) and the Northwestern Garter Snake (p. 216).

Morphology: The form, structure, and outward appearance of an organism.

Musk: In snakes, a foul-smelling substance produced by cloacal glands and released from the vent (usually in combination with feces). It acts as a defensive repellent, and in some species the musk causes irritation and pain when it comes into contact with a predator's eyes.

Naturalist: A person who studies nature, especially by direct observation out in the field.

Nocturnal: Active at night.

Oviparous: Producing eggs that hatch after laying.

Parapineal organ: A transparent disk that is light sensitive and is located on top of the head in many species of lizards. Sometimes referred to as the "parietal eye" or "third eye," it is connected by nerves to the brain through an aperture in the skull. Research indicates that this organ probably helps regulate the amount of time a lizard spends in the sun.

Parthenogenesis: The development of a female animal's ovum (egg) into an embryo without fertilization by a male. Parthenogenic species can reproduce without mating, resulting in entirely female populations. See the species account for the Plateau Striped Whiptail (p. 251) for an example.

Plastron: The underside of a turtle's shell.

Playa: The flat floor of a lakebed in a desert basin, usually dry most of the time. The surface is frequently composed of white alkali.

Postanal scales: Two or more enlarged scales situated just below the vent at the base of the tail on the males of iguanid (crotaphytid and phrynosomatid) lizards.

Predator: An animal that hunts other animals for food.

Prey: An animal eaten by a predator.

Race: See *subspecies.*

Range: The overall geographic area where a species occurs.

Rear-fanged snake: Refers to certain species of snakes that have enlarged teeth at the rear of the jaw that are often grooved. When the reptile chews on its prey, venom (harmless to humans) runs down the grooves into the wound and helps subdue the struggling intended food. See the species account for the Night Snake (p. 240) for an example.

Relict population: A remnant, disjunct population that survives from an earlier period when that particular species had a wider distribution. These isolated occurrences are often at considerable distances from their species' primary range. See the distribution maps for the Sharp-tailed Snake (p. 152) and California Mountain Kingsnake (p. 194) for examples.

Riparian: The often lushly vegetated habitat next to a stream, river, lake, pond, seepage, or other area of water.

Rostral: The large scale on the tip of a snake's snout.

Savanna: A habitat of open grasslands with scattered trees.

Scalation: The arrangement of scales on an animal.

Scutes: Any enlarged scale on a reptile, such as the plates (shields) that compose a turtle's shell.

Smooth scale: A scale that lacks a keeled ridge down the middle.

Species: Organisms that are morphologically much alike and are classified as a distinct kind. In animals, individuals of a particular species can successfully mate with one another and produce fertile young.

Spectacle: The transparent covering (eye cap) on the eye of a snake.

Subadult: A juvenile animal that is nearly adult size but is not yet sexually mature.

Subspecies: A subdivision assigned to a consistent geographical variation of a species throughout a certain section of its distribution. It is designated as the third part of an organism's scientific name: genus—species—**subspecies**. Sometimes referred to as a "race."

Taxonomy: The science of classifying organisms into groups according to their relationships with one another.

Terrestrial: Inhabiting the land.

Thermoregulation: The use of behavior and/or physiological means to achieve a constant or near-constant body temperature.

Venomous: Having the ability to inflict a poisonous bite.

Vent: The external opening of the cloaca.

Ventral: The lower (underside) surfaces of an animal.

Ventral scales: The elongated scales across a snake's belly, between the head and the anal plate.

Vertebrate: An animal with a backbone.

Viviparous: Retaining the eggs within the body until they are born as fully developed young. Sometimes referred to as "live-bearing."

Vomeronasal organs: A pair of small cavities in the roof of many reptile species' mouths that analyze scent particles and send signals to the olfactory part of the animal's brain. Also called "Jacobson's organs."

Warm-blooded: See *endotherm.*

Resources

Besides the many technical publications drawn upon in the preparation of this field guide, the titles of a number of overview reference books have been given. Several are older classics that are presently out of print, but they are available in libraries and well worth seeking out. There are also listings of suggested books that cover techniques for photographing reptiles and nature, along with some journals, monographs, and websites.

GENERAL REFERENCES

Bartlett, R. D., and A. Tennant. 2000. *Snakes of North America: Western Region.* Gulf Publishing Company, Houston, Texas.

Behler, J. L., and F. W. King. 1979. *The Audubon Society Field Guide to Reptiles and Amphibians.* Alfred A. Knopf, New York.

Brown, P. R. 1997. *A Field Guide to Snakes of California.* Gulf Publishing Company, Houston, Texas.

Carr, A. 1952. *Handbook of Turtles.* Cornell University Press, Ithaca, New York.

Cox, D. C., and W. W. Tanner. 1995. *Snakes of Utah.* Brigham Young University, Provo, Utah.

Ditmars, R. L. 1936. *The Reptiles of North America.* Doubleday and Company, Garden City, New York.

Ernst, C. H., J. E. Lovich, and R. W. Barbour. 1994. *Turtles of the United States and Canada.* Smithsonian Institution Press, Washington, D. C.

Ernst, C. H., and G. R. Zug. 1996. *Snakes in Question: The Smithsonian Answer Book.* Smithsonian Institution Press, Washington, D. C.

Gordon, K. 1939. *The Amphibia and Reptilia of Oregon.* Oregon State College Monograph, Corvallis.

Greene, H. W. 1997. *Snakes: The Evolution of Mystery in Nature.* University of California Press, Berkeley.

Gregory, P. T., and R. W. Campbell. 1984. *The Reptiles of British Columbia.* Royal British Columbia Museum, Victoria.

Halliday, T., and K. Adler. 1986. *The Encyclopedia of Reptiles and Amphibians.* Facts on File, New York.

Hammerson, G. A. 1999. *Amphibians and Reptiles in Colorado.* University Press of Colorado, Niwot.

Johnson, M. L. 1954. Reptiles of the State of Washington. *Northwest Fauna* 3:5–79. Society for Northwestern Vertebrate Biology, Olympia, Washington.

Kauffeld, C. 1995. *Snakes and Snake Hunting.* Krieger Publishing Company, Malabar, Florida.

Klauber, L. M. 1997. *Rattlesnakes: Their Habits, Life Histories, and Influence on Mankind.* 2 vols. University of California Press, Berkeley.

Koch, E. D., and C. R. Peterson. 1995. *Amphibians and Reptiles of Yellowstone and Grand Teton National Parks.* University of Utah Press, Salt Lake City.

Linder, A. D., and E. Fichter. 1977. *The Amphibians and Reptiles of Idaho.* Idaho State University Press, Pocatello.

Mattison, C. 1999. *Snake.* Dorling Kindersley, London.

Nussbaum, R. A., E. D. Brodie, Jr., and R. M. Storm. 1983. *Amphibians and Reptiles of the Pacific Northwest.* The University Press of Idaho, Moscow.

Pough, F. H., et al. 1998. *Herpetology.* Prentice-Hall, Upper Saddle River, New Jersey.

Shaw, C. E., and S. Campbell. 1974. *Snakes of the American West.* Alfred A. Knopf, New York.

Smith, H. M. 1995. *Handbook of Lizards.* Comstock Publishing Company, Ithaca.

Smith, H. M., and E. D. Brodie, Jr. 1982. *Reptiles of North America.* Golden Press, New York.

Stebbins, R. C. 1954. *Amphibians and Reptiles of Western North America.* McGraw-Hill Book Company, New York.

———. 1985. *A Field Guide to Western Reptiles and Amphibians.* Peterson Field Guide Series, Houghton Mifflin Company, Boston.

St. John, A. D. 1980. *Knowing Oregon Reptiles.* Salem Audubon Society, Oregon.

St. John, A. D., and T. Titus. 1989. *Amphibians and Reptiles of the Sunriver Area.* Sunriver Nature Center, Oregon.

Storm, R. M., and W. P. Leonard (eds.). 1995. *Reptiles of Washington and Oregon.* Seattle Audubon Society, Seattle, Washington.

Van Denburgh, J. 1922. *The Reptiles of Western North America.* 2 vols. California Academy of Sciences, San Francisco.

Wright, A. H., and A. A. Wright. 1989. *Handbook of Snakes of the United States and Canada.* 2 vols. Cornell University Press, Ithaca, New York.

Zim, H. S., and H. M. Smith. 2001. *Reptiles and Amphibians: A Guide to Familiar American Species.* Golden Nature Guide Series, Golden Press, New York.

Zug, G. R. 1993. *Herpetology: An Introductory Biology of Amphibians and Reptiles.* W. B. Saunders Company, Philadelphia.

TECHNICAL REFERENCES

Avise, J. C. 2000. *Phylogeography: The History and Formation of Species.* Harvard University Press, Cambridge, Massachusetts.

Black, J. H., and R. M. Storm. 1970. Notes on the herpetology of Grant County, Oregon. *Great Basin Naturalist,* 30(1):9–12.

Brodie, E. D., Jr., R. A. Nussbaum, and R. M. Storm. 1969. An egg-laying aggregation of five species of Oregon reptiles. *Herpetologica,* 25:223–27.

Camper, J. D., and J. R. Dixon. 1994. Geographic variation and systematics of the Striped Whipsnakes (*Masticophis taeniatus* complex). *Annals of Carnegie Museum,* 63(1):1–48.

Corn, P. S., and R. B. Bury. 1986. Morphological variation and zoogeography of Racers (*Coluber constrictor*) in the Central Rocky Mountains. *Herpetologica,* 42(2):258–64.

Croghan, S. 1982. *A Survey of Amphibians and Reptiles of Southern Baker County, Oregon.* Oregon Department of Fish and Wildlife.

Crother, B. I., et. al. 2000. *Scientific and Standard English Names of Amphibians and Reptiles of North America North of Mexico, With Comments Regarding Confidence In Our Understanding.* Society for the Study of Amphibians and Reptiles.

Engelstoft, C., and K. Ovaska. 1999. *Sharp-tailed Snake Study on the Gulf Islands and Southeastern Vancouver Island.* Ministry of Environment, Lands and Parks, Vancouver Island Regional Office, British Columbia.

Ferguson, D. E. 1954. An annotated list of the amphibians and reptiles of Union County, Oregon. *Herpetologica,* 10:149–52.

Ferguson, D. E., K. E. Payne, and R. M. Storm. 1956. The geographic distribution of the subspecies of *Pituophis catenifer* Blainville in Oregon. *Copeia,* 4:255–57.

———. 1958. Notes on the herpetology of Baker County, Oregon. *Great Basin Naturalist,* 18(2):63–65.

Fitch, H. S. 1936. Amphibians and reptiles of the Rogue River Basin, Oregon. *American Midland Naturalist,* 17:634–52.

———. 1949. Study of snake populations in Central California. *American Midland Naturalist,* 41:513–79.

Green, H. W. 1984. Taxonomic status of the Western Racer, *Coluber constrictor mormon. Journal of Herpetology,* 18:210–11.

Groves, C. 1989. Idaho's amphibians and reptiles. *Idaho Wildlife.* Idaho Department of Fish and Game.

Hirth, H. F., R. C. Pendleton, A. C. King, and T. R. Downard. 1969. Dispersal of snakes from a hibernaculum in Northwestern Utah. *Ecology,* 50:332–39.

Holland, D. C. 1994. *The Western Pond Turtle: Habitat and History.* Oregon Department of Fish and Wildlife.

Hoyer, R. F., and G. R. Stewart. 2000. Biology of the Rubber Boa (*Charina bottae*), with emphasis on *C. b. umbratica,* parts 1 and 2. *Journal of Herpetology,* 34(3):348–60.

Jackson, P. L., and A. J. Kimerling (eds.). 1993. *Atlas of the Pacific Northwest.* Oregon State University Press, Corvallis.

Johnson, M. L. 1939. *Lampropeltis zonata* (Blainville) in Washington State. *Occasional Papers,* Department of Biology, University of Puget Sound, 1:2–3.

Lavin-Murcio and K. W. Kardong. 1995. Scents related to venom and prey as cues in poststrike trailing behavior of rattlesnakes, *Crotalus virdis oreganus. Herpetologica,* 5(1):39–44.

Leonard, W. P., D. M. Darda, and K. R. McAllister. Aggregations of Sharptail Snakes (*Contia tenuis*) on the east slope of the Cascade Range in Washington State. *Northwestern Naturalist,* 77:47–49.

Leonard, W. P., and M. A. Leonard. 1998. Occurrence of the Sharptail Snake (*Contia tenuis*) at Trout Lake, Klickitat County, Washington. *Northwestern Naturalist,* 79:75–76.

Leonard, W. P., and R. C. Stebbins. 1999. Observations of antipredator tactics of the Sharp-tailed Snake (*Contia tenuis*). *Northwestern Naturalist,* 80:74–77.

Linsdale, J. M. 1940. Amphibians and reptiles in Nevada. *Proceedings of the American Academy of Arts and Sciences,* 73(8):197–257.

Llewellyn, R. L., and C. R. Peterson. 1998. *Distribution, Relative Abundance, and Habitat Associations of Amphibians and Reptiles on Craig Mountain, Idaho.* Idaho Bureau of Land Management, Technical Bulletin 98–15.

Maxell, B. A., J. K. Werner, and D. Flath. In preparation. *Herpetology in Montana.*

McAllister, K. R. 1995. Distribution of amphibians and reptiles in Washington. *Northwest Fauna,* 3:81–112.

McGuire, J. A. 1996. Phylogenetic systematics of crotaphytid lizards. *Bulletin of Carnegie Museum of Natural History,* 32:1–143.

Munger, J. C., et al. 1994. *A Survey of the Herpetofauna of Bruneau Resource Area, Boise District.* Technical Bulletin 94–7, Idaho Bureau of Land Management, Boise.

Nussbaum, R. A., and R. F. Hoyer. 1974. Geographic variation and the validity of subspecies in the Rubber Boa, *Charina bottae* (Blainville). *Northwest Science,* 48(4):219–29.

Parker, W. S., and W. S. Brown. 1974. Notes on the ecology of Regal Ringneck Snakes (*Diadophis punctatus regalis*) in Northern Utah. *Journal of Herpetology,* 8:262–63.

Pavelek, W. 1957. A distributional study of known areas inhabited by *Crotalus viridis oreganus* on the east slope of the Willamette Valley. Unpublished paper, Department of Zoology, Oregon State University, Corvallis.

Reichel, J., and Flath, D. 1995. Identification of Montana's amphibians and reptiles. *Montana Outdoors.*

Ricketts, T. H., et al. 1999. *Terrestrial Ecoregions of North America: A Conservation Assessment.* Island Press, Covelo, California.

Rodriguez-Robles, J. A., D. F. DeNardo, and R. E. Staub. 1999. Phylogeography of the California Mountain Kingsnake, *Lampropeltis zonata* (*Colubridae*). *Journal of Molecular Ecology,* 8:1923–34.

Rossman, D. A., N. B. Ford, and R. A. Seigel. 1996. *The Garter Snakes: Evolution and Ecology.* University of Oklahoma Press, Norman.

Rossman, D. A., and G. R. Stewart. 1987. Taxonomic reevaluation of *Thamnophis couchii. Occasional Papers of the Museum of Zoology,* no. 63. Louisiana State University, Baton Rouge.

Roth, J. J., B. J. Johnson, and H. M. Smith. 1989. The Western Hognose Snake, *Heterdon nasicus,* west of the Continental Divide in Colorado, and its implications. *Bulletin of the Chicago Herpetological Society,* 24(9):161–63.

Ruthven, A. G., and H. T. Gaige. 1915. The reptiles and amphibians collected in Northeastern Nevada by the Walker-Newcomb Expedition of the University of

Michigan. *Occasional Papers of the Museum of Zoology*, no. 8. University of Michigan.

Slater, J. R. 1963. Distribution of Washington reptiles. *Occasional Papers*, Department of Biology, University of Puget Sound, 24:212–33.

Smith, H. M., D. Chiszar, J. R. Staley II, and K. Tepedelen. 1994. Population relationships in the Corn Snake, *Elaphe guttata. Texas Journal of Science*, 46(3):259–92.

Smith, H. M. 1996. Further evidence of the importance of the Wyoming Corridor in herpetozoan distribution. *Bulletin of the Maryland Herpetological Society*, 32(12):28–31.

Smyth, M. 1996. *Summary Report—1995 Reptile Surveys (Sheepshead, Trout Creek, and Pueblo Mountains, Harney County, Oregon)*. Bureau of Land Management, Burns District, Oregon

St. John, A. D. 1982. *The Herpetology of Curry County, Oregon.* Oregon Department of Fish and Wildlife Technical Report 82-2-04.

———. 1982. *The Herpetology of the Wenaha Wildlife Area, Wallowa County, Oregon.* Oregon Department of Fish and Wildlife Technical Report 82-4-03.

———. 1984. *The Herpetology of Jackson and Josephine Counties, Oregon.* Oregon Department of Fish and Wildlife Technical Report 84-2-05.

———. 1984. *The Herpetology of the Upper John Day River Drainage, Oregon.* Oregon Department of Fish and Wildlife Technical Report 84-4-05.

———. 1985. *The Herpetology of the Interior Umpqua River Drainage, Douglas County, Oregon.* Oregon Department of Fish and Wildlife Technical Report 85-2-02

———. 1985. *The Herpetology of the Owyhee River Drainage, Malheur County, Oregon.* Oregon Department of Fish and Wildlife Technical Report 85-5-03.

———. 1987. *The Herpetology of the Willamette Valley, Oregon.* Oregon Department of Fish and Wildlife Technical Report 86-1-02.

———. 1987. *The Herpetology of the Oak Habitat of Southwestern Klamath County, Oregon.* Oregon Department of Fish and Wildlife Technical Report 87-3-01

———. 1994. *The Herpetofauna of the Pelton/Round Butte Project: Reregulating Reservoir Area, Jefferson County, Oregon.* Portland General Electric Company, Oregon.

Storm, R. M. 1979. *Amphibians and Reptiles of the Sheldon National Wildlife Refuge.* U. S. Fish and Wildlife Service, Nevada.

Tanner, W. W. 1967. *Contia tenuis* Baird and Girard in Continental British Columbia, Canada. *Herpetologica*, 23(4):323

Tanner, W. W., and R. B. Loomis. 1957. A taxonomic and distributional study of the western subspecies of the Milk Snake, *Lampropeltis doliata. Transactions of the Kansas Academy of Science*, 60(1):12–42.

Titus, T., and G. Rahr. 1984. *Herpetofaunal Investigation of the Lower Deschutes River Canyon.* Oregon Department of Fish and Wildlife Technical Report 83-3-01.

Vindum, J. V., and E. N. Arnold. 1997. The Northern Alligator Lizard (*Elgaria coerulea*) from Nevada. *Herpetological Review*, 28(2):100.

Welsh, H. H., Jr., and A. J. Lind. 2000. Evidence of lingual-luring by an aquatic snake. *Journal of Herpetology*, 34(1):67–74.

Whitaker, J. O., Jr., and C. Maser. 1981. Food habits of seven species of lizards from Malheur County, southeastern Oregon. *Northwest Science*, 55(3):202–8.

Williams, K. L. 1988. *Systematics and natural history of the American Milk Snake,* Lampropeltis triangulum. Milwaukee Public Museum, Wisconsin.

Young, N. J. 1999. *Observations on the Distribution of the Plateau Striped Whiptail (*Cnemidophorus velox*) in Cove Palisades State Park and Its Social Interaction with the Native Western Fence Lizard (*Sceloporus occidentalis). Oregon Parks and Recreation Department.

Zamudio, K. R., K. B. Jones, and R. H. Ward. 1997. Molecular systematics of Short-horned Lizards: biogeography and taxonomy of a widespread species complex. *Systematic Biology*, 46(2):284–305.

PHOTOGRAPHY REFERENCES

Fitzharris, T. 1997. *The Sierra Club Guide to Close-up Photography in Nature.* Sierra Club Books, San Francisco.

McDonald, J. 1994. *Designing Wildlife Photographs.* Amphoto Books, New York.

Shaw, J. 2000. *John Shaw's Nature Photography Field Guide.* Amphoto Books, New York.

West, L., and W. P. Leonard. 1997. *How to Photograph Reptiles and Amphibians.* Stackpole Books, Mechanicsburg, Pennsylvania.

REGIONAL JOURNALS AND MONOGRAPHS

Northwestern Naturalist, a tri-annual journal, and the occasional monographs titled *Northwest Fauna*, are published by the Society for Northwestern Vertebrate Biology. Subjects of a herpetological nature are frequently included and pertain specifically to the Northwest. Contact: Treasurer, SNVB, P.O. Box 61526, Vancouver, Washington 98666-1526.

Western North American Naturalist is a quarterly journal that often features herpetological articles. Contact: 290 Monte L. Bean Life Science Museum, Brigham Young University, Provo, Utah 84602

WEBSITES

American Society of Ichthyologists and Herpetologists: http://www.utexas.edu/depts/asih/

Herpetologists' League:
http://www.inhs.uiuc.edu/cbd/HL/hl.html

Partners for Amphibian and Reptile Conservation:
http://www.parcplace.org/

Society for Northwestern Vertebrate Biology:
http://www.eou.edu/snvv/

Society for the Study of Amphibians and Reptiles:
http://www.ukans.edu/~ssar/

World Wide Web Virtual Library — Herpetology:
http://cmgm.stanford.edu/~meisen/herp/

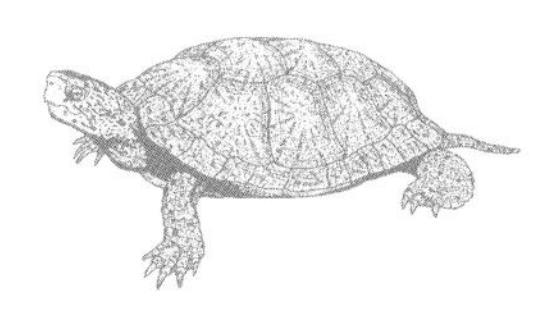

Turtles

❑ **Western Pond Turtle,** *Clemmys marmorata*
Date:
Location:
Remarks:

❑ **Painted Turtle,** *Chrysemys picta*
Date:
Location:
Remarks:

Lizards

❑ **Zebra-tailed Lizard,** *Callisaurus draconoides*
Date:
Location:
Remarks:

❑ **Great Basin Collared Lizard,** *Crotaphytus bicinctores*
Date:
Location:
Remarks:

❑ **Long-nosed Leopard Lizard,** *Gambelia wislizenii*
Date:
Location:
Remarks:

❑ **Desert Spiny Lizard,** *Sceloporus magister*
Date:
Location:
Remarks:

Western Fence Lizard, *Sceloporus occidentalis*

❑ Northwestern Fence Lizard, *S. o. occidentalis*
Date:
Location:
Remarks:

❑ Great Basin Fence Lizard, *S. o. longipes*
Date:
Location:
Remarks:

❑ **Eastern Fence Lizard,** *Sceloporus undulatus*
Date:
Location:
Remarks:

Sagebrush Lizard, *Sceloporus graciosus*

❑ Northern Sagebrush Lizard, *S. g. graciosus*
Date:
Location:
Remarks:

❑ Western Sagebrush Lizard, *S. g. gracilis*
Date:
Location:
Remarks:

Common Side-blotched Lizard, *Uta stansburiana*

❑ Typical form
Date:
Location:
Remarks:

❑ Pale variation from the Colorado Plateau
Date:
Location:
Remarks:

❑ **Ornate Tree Lizard,** *Urosaurus ornatus*
Date:
Location:
Remarks:

❑ **Desert Horned Lizard,** *Phrynosoma platyrhinos*
Date:
Location:
Remarks:

❑ **Pigmy Short-horned Lizard,** *Phrynosoma douglasi*
Date:
Location:
Remarks:

❑ **Greater Short-horned Lizard,** *Phrynosoma hernandesi*
Date:
Location:
Remarks:

❑ **Western Skink,** *Eumeces skiltonianus*
Date:
Location:
Remarks:

Western Whiptail, *Cnemidophorus tigris*
❑ Great Basin Whiptail, *C. t. tigris*
Date:
Location:
Remarks:

❑ California Whiptail, *C. t. mundus*
Date:
Location:
Remarks:

❑ Painted Desert Whiptail, *C. t. septentrionalis*
Date:
Location:
Remarks:

Southern Alligator Lizard, *Elgaria multicarinata*
❑ Oregon Alligator Lizard, *E. m. scincicauda*
Date:
Location:
Remarks:

❑ California Alligator Lizard, *E. m. multicarinata*
Date:
Location:
Remarks:

Northern Alligator Lizard, *Elgaria coerulea*
❑ Northwestern Alligator Lizard, *E. c. principis*
Date:
Location:
Remarks:

❑ Shasta Alligator Lizard, *E. c. shastensis*
Date:
Location:
Remarks:

Snakes

❑ **Rubber Boa,** *Charina bottae*
Date:
Location:
Remarks:

Sharp-tailed Snake, *Contia tenuis*

❑ Typical form
Date:
Location:
Remarks:

❑ Long-tailed form
Date:
Location:
Remarks:

Ring-necked Snake, *Diadophis punctatus*

❑ Northwestern Ring-necked Snake, *D. p. occidentalis*
Date:
Location:
Remarks:

❑ Coral-bellied Ring-necked Snake, *D. p. pulchellus*
Date:
Location:
Remarks:

❑ Regal Ring-necked Snake, *D. p. regalis*
Date:
Location:
Remarks:

- [ ] **Smooth Green Snake,** *Opheodrys vernalis*
  Date:
  Location:
  Remarks:

- [ ] **Racer,** *Coluber constrictor*
  Date:
  Location:
  Remarks:

Striped Whipsnake, *Masticophis taeniatus*

- [ ] White-striped variation
  Date:
  Location:
  Remarks:

- [ ] Yellow-striped variation from the Klamath Interior Valleys
  Date:
  Location:
  Remarks:

- [ ] Pale variation from the Colorado Plateau
  Date:
  Location:
  Remarks:

- [ ] **California Whipsnake,** *Masticophis lateralis*
  Date:
  Location:
  Remarks:

- [ ] **Coachwhip,** *Masticophis flagellum*
  Date:
  Location:
  Remarks:

Northwest Reptile Lifelist

❑ **Western Patch-nosed Snake,** *Salvadora hexalepis*
Date:
Location:
Remarks:

Gopher Snake, *Pituophis catenifer*
❑ Pacific Gopher Snake, *P. c. catenifer*
Date:
Location:
Remarks:

❑ Great Basin Gopher Snake, *P. c. deserticola*
Date:
Location:
Remarks:

❑ Bullsnake, *P. c. sayi*
Date:
Location:
Remarks:

❑ **Corn Snake,** *Elaphe guttata*
Date:
Location:
Remarks:

Common Kingsnake, *Lampropeltis getula*
❑ Black-and-white crossbanded form
Date:
Location:
Remarks:

- ☐ Brown-and-white crossbanded form
  Date:
  Location:
  Remarks:

- ☐ **California Mountain Kingsnake,** *Lampropeltis zonata*
  Date:
  Location:
  Remarks:

- ☐ **Sonoran Mountain Kingsnake,** *Lampropeltis pyromelana*
  Date:
  Location:
  Remarks:

- ☐ **Milk Snake,** *Lampropeltis triangulum*
  Date:
  Location:
  Remarks:

- ☐ **Long-nosed Snake,** *Rhinocheilus lecontei*
  Date:
  Location:
  Remarks:

Common Garter Snake, *Thamnophis sirtalis*

- ☐ Red-spotted Garter Snake, *T. s. concinnus*
  Date:
  Location:
  Remarks:

- ☐ Puget Sound Garter Snake, *T. s. pickeringii*
  Date:
  Location:
  Remarks:

❑ California Red-sided Garter Snake, *T. s. infernalis*
Date:
Location:
Remarks:

❑ Valley Garter Snake, *T. s. fitchi*
Date:
Location:
Remarks:

❑ Red-sided Garter Snake, *T. s. parietalis*
Date:
Location:
Remarks:

Northwestern Garter Snake, *Thamnophis ordinoides*

❑ Yellow-striped morph
Date:
Location:
Remarks:

❑ Red-striped morph
Date:
Location
Remarks:

❑ Orange-striped morph
Date:
Location:
Remarks:

❑ Turquoise or blue-striped morph
Date:
Location:
Remarks:

❑ Yellowish-tan single-striped morph
Date:
Location:
Remarks:

Western Terrestrial Garter Snake, *Thamnophis elegans*

❑ Mountain Garter Snake, *T. e. elegans*
Date:
Location:
Remarks:

❑ Mountain Garter Snake, Willamette Valley variation
Date:
Location:
Remarks:

❑ Coast Garter Snake, *T. e. terrestris*
Date:
Location:
Remarks:

❑ Wandering Garter Snake, *T. e. vagrans*
Date:
Location:
Remarks:

❑ Wandering Garter Snake, dark Puget Sound "*nigrescens*" variation
Date:
Location:
Remarks:

❑ Wandering Garter Snake, Klamath "*biscutatus*" variation
Date:
Location:
Remarks:

❑ Wandering Garter Snake, Colorado Plateau "*vascotanneri*" variation
Date:
Location:
Remarks:

Pacific Coast Aquatic Garter Snake, *Thamnophis atratus*

❑ Spotted morph
Date:
Location:
Remarks:

❑ Striped morph
Date:
Location:
Remarks:

❑ **Sierra Garter Snake**, *Thamnophis couchii*
Date:
Location:
Remarks:

Ground Snake, *Sonora semiannulata*

❑ Orange-and-black crossbanded morph
Date:
Location:
Remarks:

❑ White-and-black crossbanded morph
Date:
Location:
Remarks:

❑ Orange-striped morph
Date:
Location:
Remarks:

❑ Uniform tan morph
Date:
Location:
Remarks:

❑ **Night Snake,** *Hypsiglena torquata*
Date:
Location:
Remarks:

Western Rattlesnake, *Crotalus viridis*

❑ Northern Pacific Rattlesnake, *C. v. oreganus*
Date:
Location:
Remarks:

❑ Great Basin Rattlesnake, *C. v. lutosus*
Date:
Location:
Remarks:

❑ Prairie Rattlesnake, *C. v. viridis*
Date:
Location:
Remarks:

❑ Midget Faded Rattlesnake, *C. v. concolor*
Date:
Location:
Remarks:

Index

This index is of all the reptile species and subspecies mentioned in the detailed accounts (pp. 72–251). The common names of species and their primary page references are in **boldface** type to help distinguish them from the large number of subspecies names occurring in this index. The abbreviation "ssp." is used with the subspecies portion of scientific names.

About the Author

The author photographing a Great Basin Collared Lizard at Lake Owyhee State Park, southwestern Oregon (photo by Jan St. John).

A resident of Bend, Oregon, Alan St. John is a freelance interpretive naturalist who uses writing, photography, pen-and-ink drawings, and color paintings to teach about the natural world. He is the author of a book on eastern Oregon, and his work has also appeared in several Northwest regional magazines, along with national publications such as National Geographic, Ranger Rick, Natural History, Nature Conservancy, *and* The New York Times. *In the past, he has also worked as a reptile keeper at Portland's zoological park and conducted extensive reptile and amphibian field surveys for the Oregon Department of Fish and Wildlife and for the U.S. Forest Service and Bureau of Land Management.*